The Inevitable Hour

The Inevitable Hour

A History of Caring for Dying Patients in America

EMILY K. ABEL

The Johns Hopkins University Press
Baltimore

Printed in the United States of America on acid-free paper
2 4 6 8 9 7 5 3 1

The Johns Hopkins University Press
2715 North Charles Street
Baltimore, Maryland 21218-4363
www.press.jhu.edu

Library of Congress Cataloging-in-Publication Data
Abel, Emily K.
The inevitable hour : a history of caring for dying patients in America / Emily K. Abel.
p. ; cm.
Includes bibliographical references and index.
ISBN 978-1-4214-0919-1 (hardcover : alk. paper) — ISBN 1-4214-0919-4
(hardcover : alk. paper) — ISBN 978-1-4214-0920-7 (electronic) —
ISBN 1-4214-0920-8 (electronic)
I. Title.
[DNLM: 1. Terminal Care—history—United States. 2. Terminally Ill—history—United States. 3. History, 19th Century—United States. 4. History, 20th Century—United States. 5. Hospice Care—history—United States. WB 310]
616.02'9—dc23 2012035112

A catalog record for this book is available from the British Library.

Special discounts are available for bulk purchases of this book. For more information, please contact Special Sales at 410-516-6936 or specialsales@press.jhu.edu.

The Johns Hopkins University Press uses environmentally friendly book materials, including recycled text paper that is composed of at least 30 percent post-consumer waste, whenever possible.

CONTENTS

ACKNOWLEDGMENTS

Many people helped me write this book. My history writing group—Carla Bittel, Janet Farrell Brodie, Sharla Fett, Devra Weber, and Alice Wexler—read draft after draft, making generous yet sharp comments. Richard Abel, Lara Freidenfelds, Sandra Harding, Joanne Leslie, and Steven P. Wallace read individual chapters and provided extremely helpful recommendations. Jacqueline H. Wolf and Janet Golden helped me locate relevant archives.

I also appreciate the comments of audiences at the Department of the History of Medicine at the Johns Hopkins School of Medicine, the Barbara Bates Center for the Study of the History of Nursing at the University of Pennsylvania School of Nursing, and the annual conference of the American Association for the History of Medicine. I made a second trip to the Barbara Bates Center to confer with Julie Fairman and Joan Lynaugh, receiving invaluable suggestions.

Numerous archivists guided me to relevant sources, especially Steven E. Novak, Columbia University Health Sciences Library, Archives and Special Collections; Russell A. Johnson, Louise M. Darling Biomedical Library, UCLA; and Arlene Shaner, Rare Books and Manuscripts, Library, New York Academy of Medicine.

I received permissions from the State Archives Division, Kansas Historical Society, to quote from the Martha Farnsworth Collection; the Center for Oral and Public History, California State University, Fullerton, to quote from oral history interviews; the Dominican Sisters of Hawthorne Archives, to quote from the Rose Hawthorne Lathrop Papers; the Manuscript Department, Library, New York Historical Society to quote from the John Moffat Howe Diary and the Sigmund and Margaret Nestor Papers; Columbia University Augustus C. Long Health Sciences Library, Archives and Special Collections, to quote from the Presbyterian Hospital Patient Records, Elizabeth R. Pritchard, "History of Social Service of Presbyterian Hospital," and Columbia University, Office of the V.P. for Health Sciences, Central Records; the Sophia Smith Collection, Smith College, to quote from the

Dorothy Smith Dushkin Papers and the Thomas Thompson Trust Records; the *Mount Sinai School of Nursing Alumnae Newsletter* to quote from the newsletter; Special Collections Research Library, University of Chicago Library, to quote from the John Gunther Papers; the Department of Special Collections and University Archives, Stanford University, to quote from the Hawthorne Family Papers; the Schlesinger Library on the History of Women in America, Radcliffe Institute, Harvard University, to quote from the Frances Fineman Gunther Papers; the Memorial Sloan-Kettering Cancer Center, New York, to quote from a Memorial Sloan-Kettering publication at the Rockefeller Archive Center, Sleepy Hollow; and Marianne P. Brown, Santa Monica, to quote from the Parker Family Letters.

Portions of chapter 3 appeared in the *Bulletin of the History of Medicine* (2011): 29–56. It was adapted and reprinted with permission by the Johns Hopkins University Press.

Jacqueline Wehmueller, Executive Editor of the Johns Hopkins University Press, encouraged this project from the beginning and provided both trenchant and supportive comments as it progressed. Carolyn I. Moser's meticulous copyediting also improved the manuscript.

My children and grandchildren have kept me connected to happier events than those chronicled here. My husband, Rick, has shared my delight in their lives as well as so much more.

The Inevitable Hour

Introduction

"I had a very painful introduction to being a doctor," recalled Katharine Sturgis. During her first weekend as an intern in 1935, "a little girl was brought in from the housing project across the street from the hospital; she was convulsing and was unconscious." Sturgis's distress stemmed partly from her failure to maintain professional distance. The patient "was just a little bit older, maybe a year older, than my own daughter—a beautiful little girl." But Sturgis also had to confront her powerlessness in the face of death. Her disjointed sentence evokes the strength of her feelings, decades after the event. The girl died of encephalitis. "And as she was dying, I was unable—I had to run behind the screen intermittently because I was so heartbroken—I realized I was losing her." Medical school had taught Sturgis how to cure but not how to cope when her treatments proved ineffective. "I'd had every consultant I could lay my hands on. We'd done everything we could, and I couldn't save her." Although Sturgis made those remarks long after the advent of medications that might have reversed the disease outcome, she concluded, "Of course, in a way it was a very good first experience, because it certainly showed the inadequacy of medicine."[1]

Sturgis's insight has no place in medicine's triumphal epic, which emphasizes the sensational advances that have rescued people from death. Armed with new bacteriological knowledge, early-twentieth-century public health officials successfully battled the epidemic diseases that previously had ravaged nineteenth-century cities. The development of anesthesia and asepsis enabled surgeons to penetrate far inside the body, eliminating problems that had devastated life. Equally remarkable achievements in chemotherapy and radiation soon converted other fatal afflictions into curable ills. As infant mortality declined, life expectancy rose sharply, from 47 years in 1900 to 65 in 1945 and 70 in 1965.[2]

That heroic narrative has a powerful hold on the popular imagination. Americans long considered themselves a uniquely blessed people, immune to the normal vicissitudes of life. Alexis de Tocqueville famously described the country as "a course almost without limits, a field without horizon," a place where "the human spirit rushes forward . . . in every direction."[3] For more than a century, medicine has served to epitomize the country's promise, offering continual rejuvenation and renewal. Dazzled by widely heralded discoveries, we minimize the failures of medical wizardry and the complicated realities of patients' lives. And we try to forget that death eventually comes to us all.

During the past few decades, social and cultural historians have disrupted the chronicle of progress by highlighting the nonmedical factors responsible for the mortality transition as well as the enormous gulf between medicine's promise and performance. But those scholars, too, hew closely to the contours of the field they study. Because medicine's primary goal is the preservation of life, historians focus on activities directed toward that end. Although some studies follow physicians into the dissection room, few turn their gaze on the care delivered to people approaching death.[4]

This book places those people at the center of the story to ask how the metamorphosis of health care altered the experience, cultural representation, and meaning of death and dying between 1880 and 1965. As the virulence of acute infectious diseases gradually lessened, both the average age when people died and the major causes of death changed. Over the period, people with terminal illnesses began leaving home to seek care in hospitals and other institutions. Physicians' growing power and prestige helped them conceal dire diagnoses and prognoses from their patients. And many physicians began to view death less as a human norm and more as a medical defeat.

A key theme of this book is how medicine's imperative to avert death increasingly took priority over the demand to relieve pain and suffering at the end of life. "The Serenity Prayer," commonly attributed to Reinhold Niebuhr, has become almost cliché. ("God grant me the serenity to accept the things I cannot change, courage to change the things I can, and wisdom to know the difference.") But "serenity" and "courage" had very different meanings in different historical periods. Well aware that they possessed few effective treatments, nineteenth-century doctors often had little choice but to wait for the end. Popular attitudes helped patients, families, and doctors accept the inevitability of death. The dominant culture continually issued reminders of life's fragility. And dying people and their families believed they could anticipate heavenly reunions. Developments in scientific medicine at the turn of the twentieth century dramatically heightened physicians' confidence in their curative powers. Professional rewards increasingly came primarily to those who could forestall death. Rising popular expectations of medical prowess encouraged individuals to fight for recovery even when they had only a remote chance of achieving it.

This book challenges three common assumptions. One is that medicine actively sought jurisdiction over death and dying just as it did over childbirth.[5] Although death began to move into hospitals in the early twentieth century, this change occurred against rather than at the urging of those facilities. Following the example of their nineteenth-century predecessors, early-twentieth-century hospitals employed admission and discharge criteria geared toward keeping mortality rates low. The proportion of all deaths occurring in hospitals rose primarily because hospitals grew rapidly in size and number, pressure to admit dying patients overwhelmed exclusionary policies, and expectations of cure were often greatly inflated.

Another widespread assumption is that the growing role of medicine in death and dying eclipsed that of families and religion. Philippe Ariès contended that, although families did not initiate the transfer of death to hospitals, they were happy to separate themselves from the dying process.[6] The historical record demonstrates, on the contrary, that many relatives strongly resisted suggestions that they surrender fatally ill family members to medical facilities. Others traveled long distances and circumvented institutional regulations to hold extensive deathbed vigils. The theory of the secularization of society argues that confidence in scientific expertise gradually replaced faith in religious authority.[7] Many medical historians embrace that theory, but a growing literature reminds us that secularism often coexisted with religion rather than supplanting it.[8] Even during the middle decades of the twentieth century—the apogee of popular enthusiasm about medicine's power to vanquish mortality—most Americans belonged to religious congregations and believed in God and the afterlife.[9] This book demonstrates that even many people who eschewed these specific convictions relied on spiritual resources to frame the experiences of suffering that scientific medicine could not prevent.

Finally, this book undercuts the prevalent belief that most of today's problems in hospital care of dying people stem overwhelmingly from the growing dominance of intensive care units and associated technologies developed since the 1970s.[10] The truth is that long before the advent of defibrillators, feeding tubes, and respirators, dying in a hospital was an extremely dehumanizing experience. Many patients unable to pay privately were shunted to public institutions, where miserable conditions often hastened death while intensifying its agonies.

My title comes from the ninth stanza of Thomas Gray's 1750 "Elegy Written in a Country Church-Yard": "The boast of heraldry, the pomp of power / And all that beauty, all that wealth e'er gave / Awaits alike th' inevitable hour / The paths of glory lead but to the grave." William Osler, the most celebrated physician in America at the turn of the twentieth century, briefly considered calling a proposed book on death "The Inevitable Hour."[11] Osler's interest in the end of life was highly unusual. At a time when medical schools, journals, and meetings rarely discussed mortality, Osler described himself as a "student of the subject." A bibliophile, he devoted one section of his voluminous library to books on "death, heaven, and hell."[12] And in 1900, he inaugurated a study of dying, asking nurses to document the "sensations" of hospital patients close to death. Although the results were not published, Osler told his audience at a prestigious Harvard lecture in 1904 that he had "careful records of about five hundred death beds." Of these dying patients, "ninety suffered bodily pain or distress of one sort of another, eleven showed mental apprehension, two positive terror, one expressed spiritual exaltation, one bitter remorse." For the great majority, death was like birth, a "sleep and a forgetting."[13]

One reason Osler may have been able to emphasize the peacefulness of the dying is that he restricted his purview to the brief period immediately preceding death. This book examines a much longer period. Even when death arrived suddenly, as a result of accidents or acute illnesses, dying often extended over days or even weeks. In cases of "incurable" diseases, finitude was anticipated for months or years.

Although the label "incurable" was widespread in the past, it has just begun to attract historical attention. The term typically referred to diseases that were both long-term and fatal and inflicted serious pain and suffering.[14] Throughout the nineteenth century, the most fearsome one was tuberculosis (or consumption, as it frequently was called). As the mortality rate of TB waned after the turn of the century, two long-term, noninfectious ailments—heart disease and cancer—gradually surpassed it as a major cause of death. "Chronic" was closely related to "incurable." Contemporaries sometimes used the two words interchangeably, sometimes referred to diseases as both chronic and incurable, and sometimes employed "chronic" as a euphemism for "incurable." But "chronic" was also occasionally applied to stable conditions that did not cause mortality. Incurable patients always remained in the shadow of death.

If Osler's interest in the process of death distinguished him from most members of his profession, his attitude toward people with incurable disorders did not. By the early twentieth century, old age itself frequently was considered an incurable condition. In his 1905 valedictory address at Johns Hopkins, Osler again displayed his knowledge of literature, this time referring to Anthony Trollope's "charming novel," *The Fixed Period*. "The plot," he noted, "hinges upon the admirable scheme of a college into which sixty men retired for a year of contemplation before a peaceful departure by chloroform. That incalculable benefits might follow from such a scheme is apparent to any one who, like myself, is nearing the limit, and who has made a careful study of the calamities which may befall men during the seventh and eighth decades."[15] In response to the uproar those remarks incited, Osler insisted he had intended to be humorous. But it is clear that he was only half joking. Throughout his career, he had often encouraged his colleagues to withdraw from treating elderly patients because such patients faced inexorable mental and physical decline.[16] The glamour and excitement of new cures for acute diseases led other physicians to abandon incurable patients of all ages who had problems that resisted remediation.

During the early twentieth century, improved control of acute infectious diseases began to foster the illusion that medicine could triumph over death itself. Simultaneously, however, the rising prevalence of chronic conditions made any celebration premature. Between 1870 and 1920, the proportion of deaths attributed to chronic diseases increased from 7 percent to 50 percent; by 1940, the figure was more than 60 percent.

By mid-century, a few commentators began to argue that those statistics demanded a drastic reallocation of health care priorities, with far more resources being devoted to chronic diseases.[17] But even while highlighting the link between chronic disease and mortality, advocates tried to raise the status of sufferers by distinguishing them from dying people. Before the outbreak of World War II, doctors stressed that early detection combined with dramatic new treatments could save many lives. After the war, a heightened focus on rehabilitation further helped to dispel what one observer called the "neglect and pessimism" traditionally surrounding chronic diseases.[18] A New York State health official noted that, for many people, "chronic illness means only the terminal, hopeless stages of illness." Such an impression was wrong, he explained: appropriate physical and occupational therapies could restore most patients to lives of usefulness and self-sufficiency.[19] Rehabilitation also saved money. Because rehabilitation programs changed "the individual into a self-sustaining productive member of the community," they were "economically and socially sound."[20] After the 1954 passage of the Vocational Rehabilitation Act, one observer estimated that approximately two million adults who were now idle could be employed and pay taxes as a result of newly funded rehabilitation programs.[21]

Rehabilitation did not always exclude people with "hopeless" prognoses. "Rehabilitation is applicable alike to persons who may become employable and to those whose only realistic hope may be a higher level of self-care," declared a prominent commission on chronic illness in 1956.[22] Programs could teach many fatally ill patients to get in and out of bed, dress, and feed themselves with little or no assistance. Nevertheless, the overriding stress on the restoration of economic productivity strengthened the conviction that people who were too sick to return to the workforce drained public coffers and lacked social value.[23]

The increasing use of the word "terminal" to replace "incurable" did not decrease the indifference and hostility surrounding this population. Although that label typically applied to all patients who had exhausted available treatments, some commentators reserved it for people whose life expectancy was less than a year or who were too debilitated to manage on their own.[24] Regardless of the precise definition, patients deemed terminal continued to be devalued and avoided.

The evasiveness surrounding patients approaching death complicated the research for this project. Because a growing number of early-twentieth-century hospitals affiliated with medical schools emphasized anatomy, hospital records are filled with information about autopsies. Administrators repeatedly urged staff to boost rates of autopsy and recommended strategies for doing so. Doctors traded stories about their successes wresting consent from reluctant families. But medical archives say relatively little about the care of bodies before death occurred. Although I make extensive use of patient clinical records, hospital reports and regulations, and medical journals, I also rely on an array of other sources, including the biographies and personal writings of patients and families, newspaper accounts, contemporary sociological studies, chaplaincy records, and the case files of social work and charitable agencies.

If this project required more arduous detective work than anticipated, the primary sources unearthed were nevertheless vast enough to demand selectivity. I trace changes over more than a century and examine deaths from a wide range of diseases. In an attempt to avoid overgeneralizing, I chose topics that illuminate broader themes. I pay particular attention to individuals dying of cancer. Although heart disease was the major killer throughout much of the twentieth century, cancer was the most dreaded malady. Because doctors often pronounced patients with cancer incurable months and sometimes years before they died, cancer was often used as a synonym for terminal disease.[25] A discussion of the affliction enables us to explore various issues. To what extent did doctors in different periods disclose bad news to patients? To what extent did families seek to shield terminally ill loved ones from the truth? How did both patients and families respond to the enormous gulf between the promise of therapeutic breakthroughs and the reality of their own experiences? What were the codes of behavior for persons with incurable cancer? What facilities were available for them? To what extent did patients and families use the long period of advancing disease to come to terms with mortality?

Mental illness, by contrast, receives little attention in this study. I examine a wide array of medical institutions, but I do not focus on mental hospitals. Because they, too, represented important sites for death and dying, they deserve separate analysis.

To understand the religious attitudes of family members as well as of patients, I analyze the writings of bereaved relatives. The emphasis of the book, however, is on the period before death. I thus omit the growing stress on autopsies, the rise of the funeral industry, and the construction of cemeteries. Although I direct special attention to the experiences of people from socially disadvantaged groups, my discussion of the nineteenth century focuses overwhelmingly on white patients and families. Finally, the book uses illustrations from medical institutions, health professionals, and patients throughout the United States, but it emphasizes those in New York, a city with unusually well-developed health and welfare services. Dying people in other parts of the country who sought care outside the home had many fewer options.

The past explored in this book illuminates the present. Today, it has become almost commonplace to decry end-of-life care. Seeking to understand the origins of our most serious problems, many commentators draw uncritically on Philippe Ariès' pathbreaking work, which offered only fragmentary data to support sweeping generalizations.[26] Others simply juxtapose an idealized nineteenth-century death with a gruesome ending in an intensive care unit of a twenty-first-century hospital.[27] By examining the many ways in which the treatment of people approaching mortality was influenced by the transformation of medicine and health care between the late 1880s and mid-1960s, this book helps to explain both why a movement to restore dignity to the dying arose in the early 1970s and why its goals have been so difficult to achieve.

CHAPTER I

The Good Death at Home

Death was never far from Mary Ann Webber's mind as she sat by herself in her small Vermont farmhouse each evening. By the early 1860s, her five surviving children had grown and left home. Especially during the winter, when snow made roads impassable for months on end, she saw no one but Perley, her irascible and silent husband. Letters to and from her children represented virtually her sole means of communication.

In the midst of the Civil War, she recorded the names of local boys killed at the front. She also noted the deaths of many neighbors, friends, and relatives. And she was constantly aware that the symptoms of advancing years presaged her own end. A severe attack of rheumatism prompted this remark: "We know that the seeds of death are within us, therefore when in health we should often think of that inevitable doom which awaits both ourselves and our friends, so that when the summons comes which calls us or our friends away we may be prepared to meet the event with Christian fortitude."[1] Above all, she worried about both the physical and spiritual health of her children and grandchildren. One spring day she interrupted her housework to send this warning to her daughter Eliza, who had moved across the country: "Many deaths have occurred with young children in this vicinity in the past few months by scalding. Do be careful always."[2] In addition, Mary Ann continually beseeched Eliza to keep the Sabbath and read the Bible regularly: "We all of us in the world of sin and sorrow need the support of the religion of the Bible to enable us to live as we should, and life is so uncertain we cannot estimate its value to prepare us for our eternal home."[3]

Mary Ann's preoccupation with mortality was hardly unusual. Although she wrote during the most devastating carnage the nation had known, death seemed omnipresent at other times as well. Because both maternal and infant mortality rates were high, childbirth aroused special terrors. Mary Ann attended her oldest daughter, Mary Adams, when she gave birth the first time, in October 1861.

Announcing the event to her other children, Mary Ann expressed thanks that the mother was spared and then explained: "We thought Saturday morning that she would die or at least we very much feared the event, but God has saved her. She is not yet out of danger although she now is as comfortable as could be expected." The baby was "a proper healthy child, fair and bright looking."[4] Even so, he was gone before his first birthday.

The birth of subsequent children revived the sense of loss. Mary Ann did not doubt that rekindled pain overwhelmed Mary's delight in the arrival of her second son the following year. "I have thought much of you for two or three weeks past," Mary Ann wrote to Mary after his arrival. "I knew every day would be crowded with its sad memories. I have just read over your letter which brought the sad tidings of the removal of my first grandchild. It is so touching a history of its last sufferings that I thought you would like to preserve it. I felt sad to read it over and you will also, so therefore I will send it."[5]

A Minnesota woman found that an infant seen as a replacement for her dead child appeared especially vulnerable. A year after the death of Gro Svenden's young daughter Sigri in 1876, Gro gave birth to another girl, whom she named Sigri Christine. "I have a feeling, a foreboding, that this little one, who is so very dear to all of us, will not be with us long," Gro wrote to her family in Norway. "I feel that the good Lord has given her to us for just a short while. Maybe it is because she is so much like the departed one."[6] Older children's lives also seemed more precarious after death visited the family. "Cannot keep despairing now of the other children, and thinking how they will look when dead," Fanny Appleton Longfellow (Henry Wadsworth Longfellow's wife) wrote two days after her baby died. "Their gleeful voices agonize me. . . . When Death first enters a house, he throws so long a shadow—it seems to touch every one."[7] Apologizing for writing at such length about her 4½-year-old son, Jane Conine Hawkins, an Indiana woman, explained to her sister, "Remembering that one bright bud was suddenly transplanted to a more genial clime therefore it is with the greater anxiety that I watch this *little one*."[8]

In epidemics of scarlet fever, typhoid fever, diphtheria, and smallpox, more than one family member might die. Six months after the death of his infant son, physician Alexander Anderson lost his wife, brother, father, mother, mother-in-law, and sister-in-law in the yellow fever epidemic of 1798. Soon afterwards, he gave up the practice of medicine, at least partly because he recognized its impotence in the face of such disasters.[9] In February 1859 Emily Hawley, the daughter of Michigan farmers, noted in her diary that her sister was "real sick with scarlet fever. Mrs. Wiley's little girl died with it last night." The following month Emily wrote, "Not hardly a day but we hear of a death, sometimes more. There were five funerals . . . last Sunday." Five days later she reported, "Attend the funeral of Mrs. Wiley's' little boy . . . , now all their children are dead, sad, sad."[10] When the same disease struck the small Nebraska town where Sarah Jane Price taught twenty years later, she remarked, "Our little graveyard is filling up very rapidly."[11]

The obligations of neighborliness meant that serious illness and death frequently brought care and support from the entire community. The closeness of communities also reinforced the sense that death was everywhere. "I had just got prepared to take up my work," Sarah Connell Ayer wrote in her diary in October 1824, "as they sent in to me that Mrs. Appleby was dying. I hastened in, and found her in the last agonies of death, surrounded by her afflicted family." A few days later Sarah "had just sat down to breakfast," when "Miss Jones ran up, to tell me Mrs. Emery, another near neighbour, was just gone. I got up from the table and went down."[12] Mary E. Sears noted that the night before her older sister Eliza died in a small Ohio town in 1860, "many kind friends and neighbors came to render some assistance. Nothing was left undone."[13] The wife of a small farmer in Arkansas, Nannie Stillwell Jackson, frequently exchanged goods and services with at least twenty other people, both white and African American. Nannie and her husband spent the night with Mrs. Hornbuckle after her son died in 1890. When Mrs. Archdale became seriously ill on April 13, 1891, Nannie visited her half a dozen times, and she and another friend spent the night. The following day Mrs. Archdale died. Nannie, her husband, and four other neighbors all kept vigil.[14]

No professional armor protected neighbors from identifying with the troubles they witnessed. Some tragedies resonated especially deeply. Many female attendants who watched women bleed to death in childbirth were pregnant themselves. Many neighbors caring for seriously ill and dying infants in the community had children the same age at home. "Today I was to a house of mourning," Laura I. Oblinger wrote from Ottawa, Minnesota, to her husband Uriah in 1882. "Mr Gibbs' baby died yesterday it had been sick since Monday with lung fever and died yesterday 2 o'clock I was there when it died. . . . As I gazed on it as it lay stiff and cold I thought what if that was one of our little darlings."[15]

Popular culture constantly reminded people of life's fragility. The religious tracts, popular health books, and novels flooding the market insistently warned that death could come at any moment and that one must always be prepared. Children as well as adults received that message. Death and dying were dominant themes in antebellum Sunday School books. A poem in *Little Verses for Good Children* ended this way: "Lord, grant that I / In faith may die / And live with thee / Above the sky."[16]

Popular literature also prescribed how the dying should act. The concept of *ars moriendi*, or the art of dying, derived from fourteenth-century Catholic manuals.

In 1651, the Anglican bishop Jeremy Taylor published his enormously influential book *The Rule and Exercise of Holy Dying*, which both drew on and revised those texts for a Protestant audience. By the early nineteenth century, Taylor's notion of the "good death" had spread widely through American society, reaching people of various social and religious backgrounds. The basic elements of that death were consciousness and lucidity, resignation to God's will, and fortitude in the face of physical pain and emotional suffering.[17]

A major component of nineteenth-century caregiving, then, was fostering spiritual preparation. In May 1863, Mary Ann Webber's son Alpha went with two younger sisters to sell children's books door-to-door in upstate New York. The project ended abruptly when he contracted typhoid fever. Informing their parents of the disaster, the sisters wrote that they not only sat with Alpha at night and rubbed his body with alcohol throughout the day but also "read in the Bible everyday to him." In addition to a doctor, two ministers came "a number of times and . . . prayed with him." Perhaps as a result, Alpha was "perfectly prepared to die."[18]

Eight years later, Mary Ann performed that service for her husband Perley. "Your Father has been sick," she informed her children in May 1871. "Had the Doctor three times, I never left the house for more than a week, he had the most serious bilious attack I ever knew him to have. The Doctor says it is liver."[19] On June 11 he was no better. Although he still wanted to get dressed and sat up for a while every day, she had to assist him "in walking, in sitting up and lying down." The doctor "says there can be no hope of recovery. He may die very suddenly or he may live two or three weeks." "Oh," she continued, "my heart is full to see him sinking, suffering, with no power to help, and I so dread the last event."[20]

Although we cannot know why she feared that event, we do know that she and Perley tried to make sure it would be as peaceful as possible. They made allusions to their past, forgiving each other and asking God to forgive them. Perley also "offered up a short prayer, calm and with feeling, confessed the shortcomings of his life, [and] prayed earnestly for Christ's sake."[21] Four days later, he was no better. He was only "partly dressed," sat up very little, and said he was too tired to "say much on the concern of his soul." Nevertheless, the minister who prayed with her husband was convinced he gave "a comfortable evidence of his trust in Jesus."[22] In mid-August, Perley was "under the influence of morphine" and "in a stupor most of the time." His spiritual state remained Mary Ann's major consolation. She thanked the Lord that she had "been permitted to hear him offer up [his] prayers."[23]

Some patients had much longer to prepare for the end. The most common killer was tuberculosis—or consumption, as it still was called—which typically lasted several months or years and was responsible for 20 percent of all deaths.[24]

One historian writes that throughout the nineteenth century the disease was idealized as a "blessing in disguise which allowed time and mental clarity for spiritual reflection and improvement."[25] But consumption was also the most feared disease of the era, not only because it almost always concluded fatally but also because it inflicted enormous pain and suffering during the final stages. The widespread belief that positive emotions could facilitate healing occasionally encouraged family and friends to try to conceal this diagnosis. An Oregon mill worker, Henry Guernsey, initially attributed the poor health of his wife, Theisa, to dyspepsia. In November 1876, however, he noted that she had "a very bad cough" and "sometimes raises blood."[26] Hoping that dry air would help her recover, Henry moved the family to southern California, where he learned the meaning of those symptoms. In July 1877, he announced that although Theisa "thinks she has taken cold," he knew "she has the consumption. . . . I fear she will not live very long though she does not seem to think but she is improving." Henry concluded that allowing his wife to hope for survival was more important than maximizing her chances for salvation. "Please do not make any reference to what I have said in writing," he urged, "as she will be more at ease in her own belief. She some times speaks of what we would do if she would not get well but I get her mind off the subject as soon as I can for it has a tendancy to keep her down harted."[27]

Many more religious caregivers used the long period of invalidism to try to exert a pious influence. Samuella Curd had been married for two years to a Missouri merchant and had just given birth to her first child when her husband became seriously ill with consumption. As the gravity of his illness became clear, she expressed increasing despair about his religious lapses. "Oh! That Mr. Curd might be made to consider and not harden his heart," she wrote in October 1861, after hearing "a most excellent sermon to the converted." A few months later, her hopes began to be fulfilled. On January 18, 1862, she described "a talk with him upon the subject of religion" as "gratifying." The following month was even more encouraging: "Have had some satisfactory talks with him on the subject of religion and while his evidence is not as bright as I would hope yet I believe him to be a changed man. I pray God he will give him brighter manifestations." She was able to proclaim her enterprise a success on March 17. After she rubbed her husband's legs in the middle of the night, she and he "got to talking on religion to my great delight said he would join the church very soon. I believe him a genuine Christian." Five days later he was "received into the church," and on March 30 Samuella wrote: "A day long to be remembered by us all. Mr. Curd made a public profession of religion, there was a large congregation and many with whom he joined in sin, saw him make it God grant it may be blessed. Oh! It did delight my heart to see him in his weakness give himself up to God."[28]

The role of family and friends was critical at the moment of death, when they were expected to discern the state of the soul and the prospect of everlasting life.[29] Whenever possible, large numbers of onlookers gathered around the deathbed. "The room was full of *people*," wrote Martha Shaw after her husband died in Topeka, Kansas.[30] Solitary deaths were greatly feared not only because the dying lacked physical and emotional assistance but even more because no one could witness, interpret, and record the event. In 1866, when Mary Ann Webber's daughter Emma made plans to travel to join a sister in California, the oldest sister, Mary Adams, protested: "I do not think she should go without someone to take care of her. She might get sick and die on the voyage and be buried at sea without anyone to tell the story."[31]

Bystanders occasionally detected signs of eternal damnation. Mary Ann Owen Sims, an Arkansas woman, could not forget the terror she experienced at the deathbed of her uncle in 1843: "For he was a wicked man and oh the agonies of a lost soul usherd in to the presence of its Maker was truly heart rendring even to my youthful minde."[32] Fortunately, such incidents were relatively rare. Although Mary Ann Sims's husband endured "sevier" suffering, his behavior conformed to the cultural code. He exhibited "Christian fortitude and resegnation knowing that he who governed the universe done all things well." "Not a murmer escaped his lips."[33] Sarah Connell Ayer's journal entry for July 10, 1827, read: "Mrs. Carpenter died. We have reason to hope that she has exchang'd this world for a better. She suffered much pain during the last two, or three weeks of her life, but appear'd calm and resign'd to the will of her Saviour, and gave cheering evidence of her preparation to meet death."[34] Mary E. Sears described her sister's death this way:

> Eliza was unconscious to apperance much of the time, but always recognized the voice of her *dear James* [her husband], and to his questions to her readiness to go home to Jesus, she would reply in monosyllables, of ready "happy, yes, yes, happy" was Jesus calling for her? "yes, calling, calling," when a heavenly smile would light her countenance: ah! What a sweet consolation to the stricken friends to hear from her own lips the words, "ready, willing"! . . . About *four o clock* P. M. she seemed to arouse somewhat. And calling for each of us by name bade us *farewell* and *kissed* us then was soon lost. . . . God had indeed called her; and in mercy had granted her consciousness enough to recognize us for the last time on earth; for this we will praise Him, for twas indeed a consolation to hear her once more call our names so sweetly and smile so gently "don't weep for me don't weep."[35]

A Vermont woman, Mathilda Theresa Roberts, informed her husband on July 3, 1852: "Charles Stevens wife is very sick has been ever since the birth of her child in Feb. the Doctor has no hopes of her." Mathilda continued the following day: "Do not expect she will live through the day. She has given up all her family and everything of a worldly nature and longs for the time to come. Thus five children one of them a helpless infant will be left without a mother to watch over and love them. But the judge of all the Earth will do right."[36]

One response of historians to such accounts has been to argue that they bore little relation to reality. Laurence Stone, for example, asks rhetorically, "How many were so physically ravaged by pain or by disease that they were either beyond caring or a foul-smelling embarrassment to the onlookers?"[37] The linguistic turn in scholarship has sparked a different response—to understand the cultural values that shaped contemporary accounts rather than to discern the truth about how people experienced death and dying. Most notably, Drew Gilpin Faust analyzes the condolence letters penned by soldiers to the bereaved families of dead comrades. Letter after letter sought to demonstrate that the fallen had achieved the ideal of the good death not because the writers ignored contrary evidence but because they needed to "maintain the comforting assumptions about death and its meaning with which they had begun the war. In face of the profound upheaval and chaos that civil war brought to their society and to their own individual lives, Americans North and South held tenaciously to deeply rooted beliefs that would enable them to make sense out of a slaughter that was almost unbearable."[38] In far less terrible circumstances at home, contemporaries similarly searched for evidence of salvation despite the physical suffering dying people endured.

Although nineteenth-century writing emphasized the state of the soul, the body constantly demanded attention. Men and especially women administered medications to dying patients, applied poultices, watched for dangerous symptoms, changed dressings, and cleaned up vomit, excrement, pus, and blood; after death occurred, friends and family sat by bodies, washed them, and laid them out. Mary Ann Webber occasionally interrupted her account of Perley's spiritual progression to comment on his physical decline and the burdens it imposed on her. "I am holding out pretty well," she wrote several weeks before his death, "but it is hard . . . to have the care of a poor sick man day after day, week after week."[39] Despite her long experience with heavy farm work, lifting a bedridden man several times a day taxed her strength. Her expenses grew. Two years earlier, the Webbers had left their unprofitable farm and rented rooms in town. They eventually sold the farm but received far less than expected and often had to borrow to make ends meet. The cost of gin for his medicine was thus "no small item."[40] Although Mary Ann noted that the neighbors were kind, they too were poor, and at least one couple required money for their services.[41] Mary Ann had to pay Mrs. Blanchard, who dipped the extra candles that nursing required, and Mr. Blanchard, who supplied nighttime "watchers." Sickness did not soften Perley's temper. He and Mr. Blanchard frequently clashed. "Both have strong wills," Mary Ann explained to her son, "but your Father is the stronger, for he is always the conqueror. . . . Your father will always have his own peculiar traits, as we all shall, while he lives."[42]

If caregiving inflicted serious costs, however, it also provided major benefits. Above all, caregivers found satisfaction in bestowing special attention on dying relatives and friends. "I had the great pleasure of supplying all her needs and fancies," wrote Louisa May Alcott when her beloved mother was ill in 1874. After her mother's death three years later, Alcott wrote in her journal, "My only comfort is that I *could* make her last years comfortable, and lift off the burden she had carried so bravely all those years." Alcott commented in a letter to a friend, "I could not let any one else care for the dear invalid while I could lift a hand for I had always been her nurse and knew her little ways."[43] Jane Freeland, a woman in upstate New York, was gratified to discover that her presence reassured her married daughter Lelah. "It was hard for her to give up that she couldn't live," Jane wrote to her sister. "O she clung so tight to me she couldn't bear to have me out of her sight."[44]

Perhaps because so many nineteenth-century writings evaded harsh physical realities, scholars have concluded that the sights and smells of serious illness did not deeply disturb contemporary caregivers.[45] But at least some evidence suggests otherwise. Consumption's most conspicuous symptoms were emaciation and coughing. When blood vessels in the lungs broke off, people coughed up large quantities of blood. The breath smelled foul.

Popular culture reinforced the rhetorical convention encouraging writers to ignore unpleasant physical traits. The romantic heroine facing a glamorous TB death was a stock figure in nineteenth-century novels.[46] Matilda Williams wrote within that tradition when she described her daughter's last day: "I never shall forget how beautiful she looked when she commenced coughing it throwed the blush to her cheek & lips, no wax figure looked so beautiful."[47] Nevertheless, some caregivers acknowledged that advancing disease destroyed rather than enhanced appearance. Novelist and activist Eliza W. Farnham later recalled that she had never seen her sister so beautiful as she was during the early stages of consumption. "But we knew it was the beauty which ushers in decay—the rich sunset which is soon followed by blackest night. And even so it proved. The last signs of emaciation began to appear as spring passed away. When the full strength of summer came, the beauty had departed from the wasting frame, the cheeks no longer wore the hectic hue."[48] After viewing the body of a man who had died the previous night, Ellen Birdseye Wheaton exclaimed, "Poor fellow! He was dreadfully emaciated."[49]

Less privileged women were especially prone to reveal disgust. Although Jane Freeland prided herself on the care she gave her daughter at the end, Jane's initial visit was brief. "It nearly unnerved me at first the change made me sick, and had to come home for 10 days."[50] Fannie Tenney told her parents that her husband "required almost constant changing of pillows the last two or three days." Her description of his death was no more romantic: "Oh, I cannot tell you the agony of those moments, as the breathing grew harder and the film came over his eyes, the blood filling his mouth, he taking his arm from my neck to try to wipe it away, then the clinging of my hand and the last breath."[51] The wife of a Kansas farmer and itinerant preacher, Mary Abigail Chaffee, was 29 when she died. A friend who visited during her last days described her as "a complete wreck of her former life. No flesh on her bones: Never saw anyone so nervous and fretful."[52]

Many survivors' accounts appeared to move seamlessly from caring for the dying to laying out the dead, but at least some revealed what William James called the "instinctive dread" normally produced by encounter with a corpse.[53] After returning from a visit to "a relative who had been afflicted with paralysis for some months," Catharine Seely wrote: "I endeavored to prepare my mind for the impending awful event, and thought my feelings were fortified against surprise, yet when it was announced that she was gone, and I beheld the lifeless corpse, I nearly fainted—lost all power of speech, and almost of motion. Thus I see my inability to command my natural feelings, and how easily I am daunted at the sight of death."[54] Even the first sign of the end terrified Eliza W. Farnham. She recalled that the "slight dew" that "began to gather" on the forehead of her dying sister "chilled me, as if a mountain of ice had fallen beside me. It was a sign which I knew too surely. . . . I passed my hand softly over her brow. It was cold and deathlike. Then I knew the dart had gone forth. . . . Death was among us!"[55] After spending the night at a woman's house before attending her funeral, Sarah Jane Price reported that "the corpse was very offensive."[56] And women responsible for preparing the body sometimes recoiled from that task. After a neighbor's baby died, Laura Oblinger complained to her husband: "It made me very sick to work over the corpse I would much rather not have touched it. . . . There was some young ladies there that was much older than myself but they seemed to think I was a married woman & it was my place to help."[57]

Survivors also had their own spiritual work to do. Here, too, popular culture offered instruction. A vast "consolation" literature arose in the nineteenth century. Many authors were ministers, but others were women who wrote primarily for other women and typically focused on the deaths of children. Both groups urged the bereaved to accept death as the will of God and express gratitude that the patient was now free of suffering.[58] In addition, survivors were told to find comfort in knowing they could look forward to ultimate reunions in heaven. As one historian notes, "The picture of heaven as a celestial home, largely modeled on the most cherished features of the Victorian home, became widely accepted."[59] Authors tried to encourage survivors to visualize the lives of the departed in lavish detail, often engaging in activities that bore an uncanny resemblance to the ones they had left behind. In 1870, Mary Ann Webber wrote that two of her daughters had given her "a beautiful birthday gift," *The Gates Ajar*, Elizabeth Stuart Phelps's extremely popular 1868 novel, which depicted a heavenly world where soldiers conversed with Abraham Lincoln and young girls played the piano.[60]

Some letters and diaries praised survivors who remained calm and cheerful in the wake of disaster. "In my late visit to New York," Catharine Seely wrote, "I arrived at my sister's just as her little Catharine was expiring, but felt fully compensated for my exertions to be with my afflicted sister, who bore the trial with her usual fortitude and discretion, saying she could give her up not only because she was out of the reach of suffering and happy in the arms of her Creator, but she would escape innumerable trials and temptations that would probably have assailed her had she lived."[61] After visiting neighbors who had just lost their daughter, Sarah Connell Ayer reported that they "appear perfectly submissive to the Divine will."[62] Ellen Birdseye Wheaton, an abolitionist and supporter of woman's suffrage, paid a condolence call on the family of a man who had died of consumption in Syracuse, New York: "His mother is calm and cheerful, and acts as a Christian should," Ellen wrote in her diary. "I was much pleased with her."[63]

But "calm" and "cheerful" rarely were words survivors used to describe themselves. Historian Karen Halttunen demonstrates that mid-century consolation literature often encouraged displays of extravagant mourning rather than fortitude and serenity. The ultimate goal, however, remained a Christian one. The mourner "came to realize that men must not 'have a dependence on ourselves,' but must 'come out of ourselves to be able to resign themselves to God.' "[64]

Many bereaved relatives later noted that they had been too inarticulate to write in the immediate aftermath of tragedy. A Georgia teacher, for example, postponed opening her journal until six weeks after her baby's death: "I have been deferring communication with my book for weeks hoping to be able to write calmly," she explained. Grief still threatened to engulf her: "When I think of the great bereavement that has filled our heart with life long sorrow, how can I be calm! I wonder that I am sane."[65] Mrs. Olof Olsson, a Swedish immigrant in Kansas, waited a year to tell a friend of the death of her month-old infant: "I lacked so in judgment and grieved so much that I became ill as a result. I had really wanted to write to you about it, but I grieved so much that I could not gather my thoughts and write."[66]

Those who did write often expressed despair. "You have thought of me and prayed for me since you heard of our stunning calamity," Mary Wilder Foote wrote to Eliza Dwight in 1837. "I little thought on Sunday that I should ever be calm again. It seemed to me, as I bent over the cradle of my precious baby and kissed those cold still lips, that my heart would burst, and that I could not go forward one step in life, but must lie down by her side, and never, never return to the duties of life again." Mary and her husband prayed for "peace and faith" and "tried to forget our own bitter anguish and to turn our thoughts from the dreary void in life and raise them to our angel in heaven." Soon, however, "come the thronging recollections of her winning ways, her ever ready smile and kiss, her sweet voice calling for us, and we have only her empty cradle."[67] Even months later, resignation was an aspiration rather than an achievement. When a friend's son died, Mary wrote, "Oh how I longed to make her feel that in no way could she hope for peace and consolation but by realizing the spiritual existence of her child, and by bringing her life so near to his that they might know no separation. How I longed to tell her that the great object should be not to forget it in other things, but to let it hallow and consecrate the remainder of life, so that her child in heaven may be a greater blessing than he could be were he blessing her home on earth." "But," she concluded, "still more fervently do I long to *realize* all this in my own life. For it is far more easy to know what should be than to attain until it, and live it perpetually."[68]

Others similarly acknowledged that intense grief retarded efforts to fulfill pious expectations. Fannie Tenney described her response to her husband's death: "I could not look away from earth to heaven as I want to. It all seemed so terrible to me. I do say 'God's will be done.' But I cannot look up as I want to."[69] The widow of an Illinois governor, Mrs. Joseph Duncan, tried to "learn to submission" when her young son died soon after her husband, but she could not help feeling "rebellious to take away one that bore the image of his father."[70] After her daughter's death in 1837, Anne Jean Lyman, a Northampton, Massachusetts, woman, wrote to a physician, "I wish I could dispossess my mind of the weeks and months of anguish by which she was finally brought to resign this life. . . . I have found it very difficult to be resigned to her sufferings. The long and sleepless days and nights, which continued nine weeks, are ever before my imagination, like so many specters."[71]

Mary Ann Webber's daughter Mary used conventional religious rhetoric to couch her chronicle of her second baby's death. "God has again called my only son," she wrote to her mother in April 1863, "and now our little Frankie sings with his angel brother the new song of redemption." After recounting the events that occurred during the terrible days preceding his death, she described the final evening. Mary "took him on a pillow on my lap and held him till he breathed his last mortal breath. . . . In all his life he never went to sleep easier or waked up happier. I was so glad the Savior so gently took him." She urged, "Do not grieve for me, Mother. I am quite well." Nevertheless, the letter ended in a wail: "Oh, how can I live without him?"[72]

Mary Ann often had counseled her children how to face loss. Now she wrote to her daughter, "I am thankful you have taken so Christian a view of your bereavement." But she also knew that the solace of religion would not be sufficient. Perhaps she recalled her own response to the deaths of three of her own young children when she continued, "Still I am afraid you are pining your life away in secret over your sorrows."[73] Mary Ann confessed to her son, "My heart is pained within me for Mary. Her heart and affections must have been very much bound up in the life of the babe because it seemed to be given her to fill the place of the other darling baby."[74]

In a few instances, religion deepened rather than attenuated despair. Two days after John Lynds died suddenly of a sore throat, Sally Brown, a Vermont teacher, attended the funeral. "Mr. Davis preached," she wrote in her diary. "The Lynds family seemed to feel very bad and Davis seemed to rather try to aggrevate their feelings than to offer them consultation."[75] Although Sally did not record what the clergyman said, other accounts demonstrate how religious thoughts could wound rather than heal. Some bereaved parents berated themselves for having become too attached to their children. "Three oclock AM finds me watching by the cradle of our dear baby," wrote a Georgia woman. "The dreaded hour has come! The hour I have feared must come, when we should see our little one prostrated by disease—when our hearts should tremble lest the treasure be torn from our arms. Do we love her too well? Have we made her our household idol? Have we given her that place in our hearts which should be dedicated to our God?"[76] A South Carolina woman wrote to her sister: "It has pleased our heavenly father to take the dearest pledge of my affections. I had forgotten my duty and placed my whole affections on that dear child—which we are so strictly forbidden by our Bible."[77] Others feared that the intensity of grief revealed the insufficiency of faith. "This shadow of death should not rest upon our spirits if we *truly believed*," Mary Wilder Foote confided to her sister. "The bursting life around me should remind me of immortality instead of calling up sorrowing thoughts of the precious bud."[78]

According to biographer Joan Hendricks, Harriet Beecher Stowe condemned the callousness of clergymen who failed to honor the depth of mourners' anguish and thus minimized the difficulty of attaining resignation. Stowe's novels depicted "an informal 'priesthood' of women who have suffered."[79] She based that priesthood on her own life. When her fifteen-month-old son Charley died of cholera in 1849, she turned for support to a friend who had sustained a similar loss. And she used the authority derived from her experience to offer consolation to other bereaved mothers. Three years after Stowe's son Henry died from drowning, she wrote to a friend who had just lost a daughter. "Ah! Susie, I who have walked in this dark valley for now three years, what can I say to you who are entering it? One thing I can say—be not afraid and confounded if you find no apparent religious support at first. When the heartstrings are all suddenly cut, it is, I believe, a physical impossibility to feel faith or resignation, there is a revolt of the instinctive and animal system." Stowe also counseled that submission to God's will resulted from "constant painful effort" rather than "sweet attraction."[80] Nevertheless, Stowe never doubted the importance of engaging in that struggle. If she urged women to turn to other women as well as male clergy for guidance, she adhered to the basic doctrine of her religion. Only by continually trying to accept the death of loved ones as God's will could people hope to make sense of their suffering and endow it with meaning.

Others acutely aware of the gap between prescription and behavior similarly testified to the support religious teachings provided. Although Fannie Tenney failed to "look up" as much as she wished, she emphasized that "there is no other for me to go to but to God. If I did not know that there is help in him I should despair."[81] After blaming herself for having loved her daughter too well, the Georgia woman continued, "The thought of being reunited with our darling is indeed a blessed, comforting one it weakens the pang of present separation."[82] As novelist Catharine M. Sedgwick wrote to a sister whose oldest son had died, "The Holy Spirit is your comforter, and let us acknowledge the ineffable consolation with which he has softened your calamity."[83] If religion could not entirely stifle unbearable emotions, it could weaken or soften them, helping the bereaved to gain at least a small amount of distance from grief.

Mary Ann Webber's letter describing the death of her husband, Perley, is lost, but we can well imagine the information it contained. Ten years later, when Mary Ann's daughter Eliza died from pneumonia in California, Mary Adams (Mary Ann's oldest child), admonished her brother-in-law Albert this way: "I had hoped to have received letters before this telling me about her sickness, her last hours, whether she had her reason, whether she knew she might die, what she said, and all that I would have been interested in if I had been there."[84] Had she been present, Mary undoubtedly would have sat by her sister Eliza's bed, watched for signs of danger, wiped her brow, administered medication, and offered emotional solace. Mary had summoned neighbors when her own babies were dying, and we can assume she often performed those services for many others as a matter of course. Eliza had nine children, and Mary wondered what arrangements Albert had made for them. Nevertheless, Mary's primary concern was neither the physical and emotional state of her dying sister nor the care of the many children she left behind. Mary found comfort in the knowledge that "Eliza was a Christian." But the moment of death was the ultimate test, and Mary needed Albert to reassure her that Eliza had passed it well. Only by learning that she had fulfilled the requirements of a good death could Mary transform the tragedy of her sister's early demise into hope for her eternal salvation.

Medical Professionals (Sometimes) Step In

Medicine played a relatively minor role in nineteenth-century death and dying. Tuberculosis aside, most deaths arrived swiftly and suddenly, from accidents and especially from acute, infectious diseases, often before doctors arrived. The American Medical Association urged physicians not to abandon patients whose lives slowly ebbed from chronic, incurable disorders, but many continued to do so. Only the rare family could afford to hire a nurse. And the few hospitals that existed tried to keep mortality rates low, primarily by excluding patients believed most likely to die.

Doctors

The writings of family members and friends focus on the care they personally delivered to the dying. Some mention doctors, but many others explain why they had not obtained medical assistance. Some families simply could not afford a physician's fees. Without telephones or automobiles, the task of summoning doctors often involved time and effort that could not be spared in emergencies. In a small Oregon town in 1877, when Henry Guernsey's son Roy seemed to be choking, Henry tried to find someone to "go after the Dr." Because no one was available, Henry "had to take a small boat and go over alone and . . . found it a pretty hard task." As he wrote to his mother, he rowed eight miles "against the tide" to reach a doctor across the bay.[1] When some travelers reached their destinations, they found that the doctor was away, often visiting other patients. Emily Hawley, the daughter of Michigan farmers, wrote in her journal that when her sister was very sick and the family "quite afraid," her brother Henry went to get Dr. Chappell. Dr. Chappell, however, was gone, and the physician Henry next tried could not come until the next day.[2]

In many cases, doctors arrived after patients had died. After noting the various people who visited her neighbor Mrs. Archdale during the hours before her death in Arkansas, Nannie Jackson commented, "The Doctor never did come."[3] Recalling his life as a cattleman in southeast New Mexico in the 1880s, William G. Urton told an interviewer, "One could die or get well before a physician could attend any sickness on the ranch."[4]

Skepticism about physicians further deterred many people from relying on them. Children's illnesses frequently caused the greatest alarm, but they were especially likely to reveal the limits of doctors' knowledge. Many physicians did not bother to learn about children's health, assuming that mothers were competent to treat their own offspring.[5] In 1840, Rachel Simmons beseeched her mother to come and help care for a sick baby. Although a Dr. Thompson had visited and given her calomel, he "did not think he could do much for a child so young."[6] Maria D. Brown decided to avoid physicians whenever possible after her daughter died of diphtheria in Iowa in 1862. As Maria later told her daughter-in-law, "With any fair treatment, she would have pulled through. But old Dr. Farnsworth gave her terrible doses of quinine and cayenne pepper. . . . After she was gone, I said to myself, 'Never again! When the next trouble comes, it will be between me and my God. I won't have any doctor.'" Maria soon faced a "pretty hard test"—three of her children sick with scarlet fever at one time—but remained true to her resolution.[7] The divisions among practitioners helped to undermine confidence in their skills. So-called regulars were challenged by groups such as the Thomsonians in the first part of the nineteenth century and by homeopaths and eclectics in the second.

Physicians often blamed poor outcomes on delays in seeking medical attention, yet they were hardly immune to self-recrimination, especially for deaths that occurred early in their careers.[8] Many physicians were painfully aware of the deficiencies of their medical educations. A large proportion of nineteenth-century medical schools were poorly funded commercial enterprises, staffed by part-time instructors; some even lacked laboratories and libraries.[9] A 1906 survey conducted by the American Medical Association found that the quality of only half the schools could be considered acceptable.[10] Dr. J. Marion Sims, later a famous surgeon, recalled that when he graduated from the Jefferson Medical College in Philadelphia in 1837, he "had had no clinical advantages, no hospital experience, and had seen nothing at all of sickness." Summoned by a prominent community member to visit his very sick baby, Sims "had no more idea of what ailed the child, or what to do for it, than if I had never studied medicine." Even after hurrying back to his office to consult a treatise on children's diseases, he knew "no more what to prescribe for the sick babe than if I hadn't read at all." When an old nurse hired by the family informed Sims the baby would die, he expressed surprise. Soon, however, "the child stopped breathing, and I thought it a case of syncope. I never dreamed that it would die. So I jerked the baby from the bed, and held its head down, and shook it, and blew into its mouth, and tried to bring it to. I shook it again, when the old nurse laid her hand on my shoulder gently, and said, 'No use shakin' that baby any more, doctor, for that baby's dead.' "[11]

The conditions under which many doctors worked contributed to their insecurities. Charles Beneulyn Johnson, who began his medical career in rural Illinois in 1865, later wrote: "No one who has not 'been through the mill' can in any sense realize the terrible weight of responsibility that at times is thrown on the shoulders of the country practitioner. For illustration, he is called to an obstetric case ten miles in the country in the month of March with roads barely passable from a siege of rain, snow, and mildly freezing weather. The case progresses slowly, midnight has come and with its advent the patient is all at once seized with puerperal convulsions." The doctor dispatches someone to summon a consultant, who "wraps himself up warmly, gets in the saddle and starts out to make his way over the miles and miles of roads that he has to traverse with his horse. . . . " Meanwhile, "about the bedside of the suffering woman all is fear and anxiety. In one of the harder paroxysms she has bitten her lips and cheeks and now the blood is flowing from her mouth adding horror to that which the family and friends have already experienced." Although the physician administers one medicine after another, none has any effect. "As the hours pass it is noticed that she is growing weaker. Her pulse is much smaller in volume and at last is not much more than a mere flicker. Her features assume the pallor of death." The consultant arrives just after she has "taken her last breath."[12]

As Johnson indicated, the many neighbors and friends who congregated in the sickroom added to doctors' sense of unease. Perhaps partly because bystanders often assumed the right to judge medical skills and direct treatment, doctors routinely portrayed such gatherings as dangerous rather than comforting. In an account of his father's medical practice in the 1870s and 1880s, William Allen Pusey wrote: "He knew from his own experience when ill how trying were the friendliest visitors. . . . This control of visitors to the seriously sick was a difficult problem in those days." Nevertheless, Pusey added, "in very important situations," his father managed to keep neighbors "cleared out."[13]

If the site of medical care intensified the insecurities of physicians, it also encouraged them to view their patients as distinct individuals. Doctors commonly ate meals at the homes they visited. When death seemed imminent, some remained through the night. Reigning medical beliefs also discouraged objectification. Because disease was assumed to arise from the particular interaction of individuals with their environment, universalistic knowledge of physiological processes was considered less important than personal knowledge of patients and the contexts of their lives. Doctors not only had to respect family members' intimate understanding of patients but also to seek such an understanding themselves.[14]

In the absence of standardized case forms, physicians frequently recorded singular aspects of illness. In 1859, for example, Charles A. Hentz, a southern physician wrote: "Hiram Roberts died this morning near Greenwood—of pneumonia—; dissipation had left him little constitutional *vim*—; poor fellow—; he lived for *naught*."[15] More admiring of a patient he saw in 1860, Hentz took time to attend the funeral and noted the consequences of the death for others. He first visited Abner Gregory at midnight on September 3, finding him "*very* sick indeed." Two days later, he wrote, "Abner was asleep when I was there—a good sign—God grant he may recover." But on September 6, Abner was no better. Hentz's entry for September 8 read: "I attended with Dr. Telfair, the funeral of Abner Gregory today—Dr. White made a very excellent sermon at the church (in the country)—after which the body was placed in the family enclosure—Poor fellow—one of the finest young men in the country—he was engaged & soon to be married to Ann Love—; the poor girl was there, & wept as if her heart would break."[16]

Although acute illness was the primary cause of mortality throughout the nineteenth century, the few ethical writings by physicians focused on deaths that had been long anticipated from chronic, incurable conditions. Hippocrates had encouraged doctors to withdraw from patients with hopeless prognoses, leaving them to the ministrations of family and clergy, and for centuries physicians had followed that recommendation. During the late 1700s and early 1800s, however, physicians on both sides of the Atlantic increasingly cared for people with chronic diseases who were approaching death.[17] The American Medical Association's first code of medical ethics, modeled on the 1803 treatise of the English physician Thomas Percival, declared in 1847: "A physician ought not to abandon a patient because the case is deemed incurable, for his attendance may continue to be useful to the patient, and comforting to the relatives around him, even to the last period of a fatal malady."[18]

In the case of dying patients the code assigned two major tasks to physicians. The first was administering pain relief.[19] Although doctors had long had access to various pain remedies, the 1816 isolation of morphine, the discovery of ether and chloroform in the late 1840s, and the development of the hypodermic syringe in the early 1850s produced far more powerful analgesics.[20] The night before Mary Richardson Walker's husband died, a doctor administered opiates "until Mr W got easier."[21] Describing her mother's final day in 1877, Louisa May Alcott wrote that "each breath was a pain & every hour an increasing burden. Dr. Conrad Wesselhoeft & our own good Dr Cook did much to mitigate the suffering."[22]

Because the development of pain medication for death lagged far behind that for surgery, many dying people continued to experience excruciating pain.[23] Biographer John Matteson notes that although "a painless serenity attends the last days of Beth March in *Little Women*," Louisa May Alcott's sister Lizzie "was not so fortunate."[24] As Alcott later wrote, Lizzie spent her last day "in great pain begging us to help & stretching her poor thin hands for the ether which had lost its effect upon her."[25] In her description of her older sister's death from a tumor in 1860, Mary Sears noted that the doctor, "kind and attentive called often through the day and did all in his power to allay her pains and distress: but relief was only temporary."[26]

The nineteenth century emphasized spiritual preparation for mortality, so it might be expected that doctors who attended dying patients assumed that their second job was warning that the end was near. The AMA code, however, declared otherwise. The physician should avoid making "gloomy prognostications."[27] In his 1895 commentary on the code, the prominent physician Austin Flint explained that emotions exerted a powerful impact on disease outcome: "Undue solemnity, anxiety, and apprehension in the looks, manner, or words of a medical attendant on the sick, are extremely unfortunate—they discourage patients, whereas, on the other hand, a cheerful mien, calmness of deportment, and verbal assurances, sometimes accomplish more than drugs."[28] The author of the only U.S. book on medical ethics in the nineteenth century, Worthington Hooker, urged physicians to take extra care to maintain the unusually strong "tendency to hope" of people who had tuberculosis. Although such hope often was "delusive," it also "in very many cases . . . manifestly prolongs life, and adds to its comfort and usefulness."[29]

The difficulty of predicting the future also bolstered the AMA recommendation. As an 1883 editorial in the *Boston Medical and Surgical Journal* stated, "No two patients have the same constitutional or mental proclivities. No two instances of typhoid fever or of any other disease are precisely alike."[30] Because diseases did not follow the same course in every instance, those that initially appeared fatal to the physician might instead end far more happily.

The injunction to withhold bad news was not absolute. Flint argued that patients who explicitly asked if their lives were in danger needed to be answered honestly. In addition, it was important to give timely warnings to friends and family, lest they be "taken by surprise" and blame the physician for concealment. Nevertheless, Flint agreed that physicians should hide dire prognoses from patients whenever possible.[31]

Ethical codes, of course, describe ideal rather than actual behavior. The gap between the AMA's prescription and physician actions may have been especially great in the nineteenth century because the association could claim the allegiance of only a tiny fraction of the profession. We cannot determine how often hopeless diagnoses were revealed or what proportion of doctors attended patients with long-term, fatal conditions. Austin Flint may have reflected not only on the experiences of his colleagues but also on his own long and distinguished medical career when he acknowledged in 1895, "It is trying to a physician to continue to visit patients when he feels the resources of medicine are powerless, and to witness the closing scenes of life."

Nevertheless, the code helps to rebut the assertion of one physician-sociologist that prognosis "had a prominent place in medicine" through the turn of the twentieth century.[32] At least one major group of physicians was instructed to avoid making "gloomy" predictions much earlier.

Nurses

Although doctors figured prominently in accounts of nineteenth-century death and dying, nurses rarely appeared. Before nursing schools were established, in 1873, many women declared or "professed" themselves nurses. Many families could not afford what it cost to employ a nurse. Those who could afford to pay for nursing care often chose not to. Some were not used to depending on strangers for care and were fearful of allowing them into their homes. Describing her husband's death in 1881, Caroline Clapp Briggs wrote, "We were arranging for a nurse Saturday afternoon, and were dreading to put him into any hands less loving than our own, when, suddenly . . . his soul was freed from his worn and weary body."[33] Class prejudices may have heightened anxieties about employing outsiders. Whereas most families employing nurses were wealthy, nurses came overwhelmingly from less privileged backgrounds. Louisa May Alcott claimed that she intended to hire a nurse during her mother's long illness but could find none who met her standards. She later employed a nurse to care for her dying father, but only because her own poor health prevented her from caring for him herself.[34]

Family members and friends who did hire nurses escaped some of the overwhelming exhaustion of caring for seriously ill people. "You must not feel uneasy about me," Mary Serena Eliza Blair wrote to her daughter in 1875. Although Mary went every day to tend a friend near death, she explained that "an excellent nurse" performed the most onerous tasks.[35] After her mother's final illness, Ralph Waldo Emerson's daughter Ellen wrote, "It was not hard to take care of Mother this last year, at least all the hardness came on Miss Leavitt, not on me, and I am not in the least tired." Although Ellen may have underestimated her own contributions, her burdens were definitely lighter than those of many daughters with dying parents.

The work lifted from family and friends, however, fell heavily on the nurses. After becoming a professed nurse in Ann Arbor, Michigan, in 1888, Emily Jane Green Hollister cared for several critically ill and dying people. The first was Mrs. Hennequine, who had been "an invalid for years with spinal trouble" and was now "entirely helpless." Ten days after Emily arrived, Mrs. Hennequine became unconscious. She died a few days later. "The work and anxiety for those days and all the time I stayed were trying," Emily noted in her diary. In August 1890, Emily went to nurse Donald McIntrye, who was "sick and ailing with softening of the brain"; his wife was exhausted from caring for him and needed rest. But Emily, too, found him "a deal of trouble; has to be watched all the time." Another "great care" was Mr. Oswold, who was "sick with paralysis" in May 1894. A week before his death, she reported: "I care for him day and night. I get dreadfully tired."[36]

Living in patients' homes for days and sometimes weeks or even months at a time, nurses broadened their gaze far beyond the diseases they tended. Emily noted that Mrs. Prudens, an old woman with dropsy, was "very hard to please but very anxious to live." Because her doctor "gave her no encouragement," Mrs. Prudens sent for another, who agreed the end was near. Although Emily did not note any spiritual preparations, she reported that "finally" Mrs. Prudens "began to make some disposal of her goods." She gave a card and comb to her nephews, her flax wheel to a church, her feather bed ("with a blue patch on it") to her husband, and a dollar "to a cousin that she owed it to for a number of years." The house was old and neglected and reminded Hollister of "one of the old mansions with many rooms, that one reads of in books" and "gives one a creepy feeling." After the next patient, Mrs. George Lowry, died from "puerperal convulsions," Emily remarked: "A lovely young mother. There was much sadness for all." Mrs. Hennequine's illness prompted Emily to reflect on her own life: "She has a kind husband, a lovely child, a good home, but alas no health. I would not change my health for all the wealth, much as I long for more and a home."[37]

Hospitals

In hospitals as at home, patients received frequent admonitions to contemplate and prepare for mortality. Even nonsectarian facilities viewed themselves as religious and social welfare institutions as well as medical ones and emphasized the importance of spiritual and moral uplift, especially when death approached.

The Methodist hospital chaplain John Moffat Howe recorded in his diary the admonitions he delivered. In January 1838 he reflected that during the past year he had "stood by the bed side of some 40 or 50 sick & dying individuals" at New York

Hospital, the city's first privately operated facility.[38] The previous October he had described a young Englishman who had worked in a "dyeing establishment where there was a cistern of boiling water." Contrary to instructions, the man had "often laid a board across it & walked across it." The previous week he had fallen in and become "badly scalded." "When I entered the room & saw them dressing his wounds," Howe wrote, "I said to myself he must die there is no alternative but thought I—I will visit him every day to try to lead him to the Savior." Hurrying to the ward the following day, he learned the Englishman had died. Howe then turned his attention elsewhere. "I walked slowly & thoughtfully to the room in which he had lain. There was the bedstead in which he had lain but his cover over the bed & clothes was taken away. I looked around upon the men in the ward (some 10 Catholics)—and said the young man is gone—yes'm they replied. I said I did not think he would die so soon. I then remarked death usually comes unexpectedly. I exhorted them to prepare to be unexpectedly called to die."[39]

Frequent rebuffs failed to deflect Howe from his mission. In November 1837 he recorded his exchanges with an elderly Frenchman, another Catholic. "When I first came to visit him he said that if I was going to talk upon religion he would leave the room. He added that he was near death & wished to die in the religion of his fathers. I told him that I had not come to proselyte but symply to encourage him to trust in his God to serve him & that his God would not forsake him in his affliction."[40]

Other patients refused to resign themselves to the inevitable. At the beginning of April 1838, Howe "hastened" to see a young German. Although he was "in the last stages of consumption," he "appeared to be unreconciled to death." Howe thus "spoke of the sufferings of life, the afflictions etc. of the present life & the gain of dying." The German replied, "Oh, yes . . . , but it is hard for a young man not turned 30 years to die." Nevertheless, he eventually agreed that Howe could return regularly to pray with him.[41] An "Irishman by birth" also initially rejected Howe's ministrations. The man had arrived at the hospital with lockjaw (tetanus), which would probably lead to his death. "His frame is robust," Howe observed, "but he is much convulsed—he has spasms successively—he lays on his bed upon his stomach with each hand clenched to the bed stead & every few minutes grones." When Howe "conversed with him about his prospects for Death, he listened but said his agony was so great he could take but little good of it." He too, however, consented to Howe's return and "appeared to enter into prayer."[42] Howe's final visit that month ran more smoothly. A "Coloured man" who was "very near the end" furnished a model for others. Because he "clearly had accepted God," Howe asked several questions, explaining that he wanted the other men in the ward to "hear from your dying lips the goodness of God to your soul & thereby be constrained to seek after the Lord." Howe and the patient then prayed together. "It was a solemn occasion," Howe concluded.[43]

Institutions with religious affiliations were especially likely to stress spiritual preparation for death. The 1866 annual report of St. Luke's Hospital, an Episcopal facility in New York, noted: "Among the homeless and friendless sick admitted into our wards, there have been, as usual a large number who were in the last stages of irremediable disease. . . . To such sufferers St. Luke's continues to be an asylum, soothing and comforting the wasting remnant of their days, and often, happily preparing them for their final change."[44]

The Protestant ethos of both denominational and nondenominational hospitals spurred Catholics and Jews to establish their own facilities. As Howe noted, the majority of his early-nineteenth-century immigrant patients were Catholic. Despite his reassurance that he did not "proselyte," there were many complaints that Protestant chaplains encouraged deathbed conversions. The Jewish immigrants from eastern and southern Europe who filled hospital beds later in the century lodged similar protests.[45]

Because Catholics believed that the final sacraments could confer a state of grace, the women's religious orders that founded and administered the majority of Catholic facilities placed special emphasis on the spiritual care of dying patients. Viewing medical and religious work as intertwined, the sisters sought simultaneously to restore patient health and help the very sick achieve good deaths, by exhorting them in the words of two historians, "to penance, resignation, and prayer."[46]

If spiritual preparation for mortality represented an essential feature of much nineteenth-century hospital care, however, hospitals were not central to the dying experience. As historian Charles E. Rosenberg writes, "Most Americans in 1800 had probably heard that such things as hospitals existed, but only a minority would have ever had occasion to see one."[47] The situation had not changed greatly seventy years later. When the first government survey was conducted in 1873, the nation had only 120 hospitals, most of which were custodial institutions serving the "deserving poor."[48] Middle-class patients rarely entered hospitals. Although low-income people had fewer options, most families were reluctant to entrust dying relatives to such facilities.[49]

And some hospitals tried to discourage dying patients from entering their institution. Rather than the large behemoths that exist today, most hospitals were located in small homes and thus had to consider the sensibilities of neighbors who feared close association with death.[50] Superintendents also sought to allay the anxieties of potential patients who had heard rumors that doctors experimented on the living, administered "black bottles" to hasten death, and dissected the bodies of the dead.[51]

Annual reports prominently displayed hospitals' low mortality rates (typically less than 10 percent).[52] "The death ratio is a thing about which competing institutions wrangle a great deal," wrote Amos G. Warner in his widely cited book on American charities. Some hospitals calculated the rate as the proportion of deaths "to the whole number disposed of," which Warner considered the fair method. Others used "the whole number of patients" and still others the "number of days' service rendered." But even when administrators produced the death rate as Warner suggested, he noted that "its significance is still ambiguous, for it may be kept down by refusing to receive all cases where the prognosis is death."[53]

Most hospitals excluded patients with incurable diseases, especially from free beds. In a few cases, the term "incurable" meant that death was imminent. Noting that "some patients had died soon after admission," Pennsylvania Hospital physicians urged physicians "to pay strict attention to rules that say no incurable patients can be admitted."[54] More commonly, the word referred to diseases that were both long-term and fatal, such as tuberculosis and cancer. The exclusion of these patients stemmed partly from the nature of their symptoms. Tuberculosis declared itself not only in extreme emaciation but also in sounds. As Susan Sontag wrote, the "prototypical" TB sign was coughing: "The sufferer is wracked by coughs, then sinks back, recovers breath, breathes normally; then coughs again."[55] Because cancers of the face, mouth, breast, and skin typically were not treated early, many patients with cancer had open sores. There were no antibiotics yet, so sores easily became infected; many emitted foul odors. When a physician at Woman's Hospital in New York admitted two cases of cancer in 1873, the Board of Lady Supervisors protested that "the ward was made so very offensive from the nature of the disease that two or three patients actually left and others complained bitterly of the discomfort to which they had been subjected."[56] The prohibition against admitting patients with TB and cancer minimized the likelihood of such complaints while helping to reduce mortality rates at a time when people with either disease were unlikely to be cured.

Discharge policies also sought to keep the death toll low. The 1798 rules of Pennsylvania Hospital read, "All persons who have been admitted into the Hospital, shall be discharged as soon as they are cured, or after a reasonable time of Tryal, are judged incurable."[57] In 1871 Brooklyn Homeopathic Hospital regulations instructed administrators "on every visiting day" to "inquire of physicians and surgeons . . . whether any of the charity patients are incurable" and to "direct all such to be discharged, so that no improper persons be permitted to remain."[58]

It is likely that encouraging patients to die at home was often a means of lowering the death rate as well as a humanitarian gesture.[59] Hospital staff may well have disregarded exclusionary admission and discharge policies, but the intention of the rules was clear.[60]

Despite attempts to reduce the size of the dying population, hospital patients may well have had the sense that death surrounded them. The average length of stay was a month or longer, and wards contained as many as twenty-five beds.[61] As a result, most people had frequent encounters with death. Because little space separated the beds, many patients may have lain in close proximity to the dying. Moreover, deaths in hospitals, like those at home, were public events. Only the rare nineteenth-century facility furnished screens to protect the privacy of dying patients or the sensitivity of others; virtually none moved moribund patients to separate rooms.[62] Other patients thus could not escape the sights and sounds of death. One hospitalized woman complained of having to listen to the "dreadful agony" of another patient throughout the night.[63] Chaplains like Howe, as we have seen, ensured that ward patients took careful note even of deaths that occurred relatively unobtrusively. And work requirements brought ambulatory patients into contact with mortality. Some nursed sicker patients.[64] As late as 1892, the Executive Committee of Montefiore Hospital in New York instructed the superintendent "to designate such male inmates as are able and willing to wash and watch over dead male patients for the purpose of saving the expense of a hired watcher."[65]

If the lack of separate rooms made privacy impossible, it also meant that most patients did not die alone. Moreover, hospital regulations specified that doctors and nurses post "danger lists," including the names of critically ill patients, and summon their family members and clergy. Although many people entered hospitals because they had no one to care for them, some dying patients had relatives and friends who could be called. After visiting a dying Irishman, John Moffat Howe noted that the patient's "poor wife sits by his side performing all the sad offices she can."[66]

Beds for Indigent Incurable Patients

When the Board of Lady Supervisors of Woman's Hospital in New York instituted a policy refusing the admission of cancer patients, one physician reportedly protested that the board had turned "a dying woman into the streets."[67] Although the physician may have overstated his case in anger, he highlighted the plight of many people with incurable chronic illnesses, especially those who lacked substantial financial resources.

To some extent, public hospitals served as repositories for such patients. But municipal hospitals had their own eligibility criteria. Although an early plan specified that Boston City Hospital would provide beds for people who had tuberculosis as well as for other "chronic incurables," speeches at the 1864 dedication ceremony made it clear that the new facility had no such intention.[68] When a coroner condemned Bellevue Hospital (New York City's primary public acute-care facility) for sending too many dying patients to facilities on Blackwell's Island in the East River in 1855, Bellevue officials responded by pointing to their policy excluding people with incurable diseases.[69]

Two island facilities received the patients Bellevue rejected: the Penitentiary Hospital and the almshouse. Established in 1849, the Penitentiary Hospital was separated from the penitentiary eight years later and renamed Island Hospital. (It would soon become first Charity and then, in 1892, City Hospital.) An 1878 report stated, "The nature of the cases received in many of our wards is such that the wards are always full, the diseases being for the most part chronic and incurable. These miserable creatures usually enter the hospital only to die or to recuperate sufficiently to enable them to struggle for an existence a short time outside, returning after a longer or shorter time to the hospital either to die or to repeat the same experience."[70] The *New York Times* reported that during the hospital's first decades of existence, "the name 'Charity Hospital' inspired a feeling akin to terror. For it to be said that a sick person must go to 'the island' was to his friends like the seal of doom."[71] The wards were so crowded that patients often slept two to a bed. The food was poor. Workhouse inmates did much of the nursing. Horror stories occasionally surfaced about abusive care. "The Blackwell's Island Outrage," read a *Times* headline in May 1867, in an article reporting that an orderly of Charity Hospital had been "charged with inhuman treatment toward a dying inmate." According to a witness, the orderly had tied up the patient's mouth and closed his eyes while he still breathed.[72]

The other facility was the almshouse. After noting that patients transported from Bellevue represented 15 percent of the almshouse mortality in 1857, the resident physician wrote, "I can see no reason for making the Alms House a 'Dead House' for other institutions, nor can I conceive myself that it is merciful to transfer persons, in or near a dying condition, from one institution to another."[73] ("Dead house" was a common term for a morgue.) By the second half of the nineteenth century, seriously ill people represented a very high proportion of the almshouse population throughout the country. Many patients died without medical attention.[74]

In 1877, the Blackwell's Island almshouse was one of the first hospitals to establish separate wards. The facility's report that year remarked, "These have already demonstrated the wisdom of this action in the better care received by those transferred to them and in relieving these old people from the dangers and discomforts of a transfer to Charity Hospital, when, perhaps, within a few hours of death."[75] The word "care," however, may have been a misnomer. Homer Folks, commissioner of public charities, later described his first visit to the hospital wing, in the early 1890s:

> The men's wards contained some sixty patients—many of them confined to their beds, and some of them in the last stages of consumption. When the visitors inquired for the person in charge, there was considerable confusion and loud calls for, we will say, "Billy," who soon came tottering down from the upper end of the ward—a picture of confusion and incompetency. This one old man who might more properly have been a patient, had sole charge of these sixty sick people. It was a hot day in mid-summer. One poor old man who had pulled down the mosquito-netting from over his face, and was too weak to replace it, was being tortured by flies. The visitor asked the orderly if this man were not in a dying condition. The orderly, who had been paying no attention to the patient at that time, looked at him and said he "guess he was." The patient died later that afternoon.[76]

The almshouse also had a separate "incurable hospital." The 1871 report claimed that "the orderly and nurse with willing assistants" were "unceasing in their efforts to attend to the wants and promote the comfort of the inmates." The examining physician added that many patients "evince a stubborn reluctance" to enter.[77]

As the almshouse physician noted, the trip to the island was "not merciful" for people close to death. Although Bellevue Hospital attendants transported patients to the dock, workhouse inmates were responsible for helping them on and off the boat. According to Homer Folks, "the 'stretcher cases' were placed in the long-open passageways" and "exposed to the wind and cold, and protected only by a blanket" during the half-hour trip across the river. "There were no nurses to observe their condition and to supply food or stimulants if such were needed."[78] In 1869, the Committee of Inspection of the Medical Board of Charity Hospital noted that a physician had reported the "death of a fever patient from pneumonia, induced, as he believes, by being transferred from the city to Blackwell's Island in an open boat." Such cases, the committee emphasized, were common.[79] Twenty years later, Charity Hospital officials complained: "During the year, many patients were brought here in a dying condition. Their transfer, through exposure to wind and weather on a stretcher, from Bellevue Hospital, serves to increase our mortality."[80]

The Nineteenth Century as Prologue

The following chapters trace the changes wrought by the transformation of health care between the late 1800s and the mid-1960s. As spectacular bacteriological discoveries heightened confidence in medicine's ability to avert death, doctors labeled fewer and fewer conditions incurable. Chronic, degenerative diseases replaced acute, infectious ones as the major causes of mortality; as a result, death occurred later in life. Growing numbers of dying patients moved into medical institutions, where death became both more private and less personal.

Nevertheless, significant features of the nineteenth century survived well into the twentieth. Many families still demanded the right to preside over death and continued to view it as a spiritual and social event as well as a medical one. Doctors employed many of the same arguments to justify concealing bad news. And hospitals continued to try to exclude dying patients. Those from disadvantaged groups were especially likely to be considered pariahs. Private hospitals continued to dump the sickest and poorest patients on municipal hospitals. Public hospitals, in turn, dispatched that population as soon as possible to facilities which failed to provide even the most rudimentary care.

Cultivating Detachment, Sidetracking Care

The growing faith in medical science during the late nineteenth and early twentieth centuries altered the meaning and experience of death in America. An accumulation of breakthroughs, including the isolation of the pathogens causing major infectious diseases and the development of dramatic new diagnostic technologies, both enhanced medicine's efficacy and dazzled the public. Growing numbers of dying patients sought care from physicians and hospitals, but as those providers gained increased confidence in their ability to avert mortality, concerns about the physical, emotional, and spiritual sufferings of people at the end of life receded into the background. The idealization of scientific rationality also helped to alter public attitudes about death. Emboldened by the new bacteriological knowledge, health officials launched widespread campaigns to convince the public that mortality resulted from human action rather than divine forces. Rather than winning praise, patients and relatives who accepted death as God's will increasingly met criticism.

Reorganizing the Health Care System

Various health providers gained social legitimacy by allying themselves with scientific breakthroughs. Institutions revised surgical procedures to conform with discoveries about asepsis and installed X-ray machines and clinical laboratories, which publicity photographs prominently displayed.[1] As hospitals increasingly converted themselves into major scientific enterprises, growing numbers of physicians sought to affiliate with them, participate in their governance, and fill their beds with patients. A nationwide survey conducted in 1909 found 4,359 hospitals with a total of 421,065 beds.[2]

So-called "regular" physicians capitalized on developments in laboratory science to assert their dominance over competing schools. They reorganized the American Medical Association by linking it more closely with local and state medical societies; by 1910, over half of all regular physicians had become members. That newly reconstituted body helped to launch a movement for the reform of medical education, leading to the closure of large numbers of inadequate schools as well as the upgrading of surviving ones. As a result, entry into medicine was restricted to those who had completed a standardized university curriculum based on the principles of scientific medicine.[3] Although nurses never achieved equal power or prestige, they, too, sought professional status by donning the mantle of science. The first nursing schools opened in 1873, and their numbers grew rapidly, reaching 1,600 by the outbreak of World War I.[4] As late as 1920, approximately three-fourths of the graduates worked in private duty, either in patient homes or hospital rooms.[5]

The "Fatal Mistake of Caring"

"Inasmuch as all of our patients, as well as we ourselves, must die sooner or later," wrote physician Alfred Worcester in 1929, "we might naturally suppose that the care of the dying would receive more attention." Nevertheless, "in the onward progress of medical science it appears to have been sidetracked. Those who are interested only in the diseases of their patients find little that is noteworthy in their dying beyond the mere fact of fatality and the possible opportunity of verifying their diagnoses."[6] Other observers noted that discussions about care for the dying rarely appeared in medical schools, meetings, or journals.[7]

Although Worcester wrote little about nurses, their lack of attention to the care for patients at the end of life might have seemed even more surprising to him. Nurses typically had the primary responsibility for tending the dying, but their schools, conferences, and journals, too, rarely mentioned the needs of individuals close to death.[8] A 1908 textbook entitled *Practice of Medicine for Nurses*, for example, included only the following brief paragraph, focusing on nursing tasks after the event had occurred: "In Case of Death, the nurse should first notify the physician if he happen to be not present. Then she should gently lead the family from the room in order to cleanse the body and put the room in order."[9]

Despite the neglect of the dying in the curriculum of medical and nursing schools, both transmitted attitudes that significantly altered professionals' relationship with that patient population. We saw that nineteenth-century medical beliefs had encouraged doctors to value personal ties both to alleviate stress and as a source of medical knowledge. Discovery of the bacteriologic causes of specific diseases encouraged medical educators to emphasize the importance of maintaining emotional distance from patients, not of developing intimate bonds with them. Biological reductionism rendered irrelevant the patient's emotional and moral state, interaction with providers, and physical surroundings.[10] The shift to hospitals as the site of medical training strengthened that lesson. "In the hospital," historian Kenneth M. Ludmerer writes, "the power of patients was diminished, as they were detached from their home, place of work, friends, and family. These unusual circumstances, and the fact that contact with hospital personnel usually ended with discharge, made it difficult for even the most conscientious student or house officer to develop a long-term relationship or acquire an understanding of the whole person."[11] Students learned to write terse case notes in place of the rich narratives previous generations of physicians had penned.

Medical schools also encouraged students to shield themselves against patient suffering. When a physician at New York's Roosevelt Hospital lost a "desperately sick little girl of seven or eight," he "grieved as though" he "had lost a new relative." But then, "pausing to take stock of the situation," he continued, "I absorbed the hard lesson that every doctor must learn. I saw that if a physician is to perform his services with practical and effective sympathy, if he is to practice his profession wisely, efficiently and with peace of mind he must avoid becoming too involved too closely, in an emotional way, with the distress of his patients."[12]

Growing evasiveness about serious illness both mirrored and exaggerated the distance between physicians and dying patients. Doctors, of course, had long concealed unpleasant news, but now the deepening chasm between professional and lay knowledge facilitated that practice. Although little had distinguished the ideas and practices of physicians from those of laypeople throughout much of the nineteenth century, physicians could claim unique competence by the early twentieth century. They alone had access to diagnostic tools, and they spoke a language few patients could understand. Doctors also had a better understanding of diseases as distinct entities in all patients and could more accurately predict the outcome.[13] As a result, historian Robert Aronowitz notes, "the physicians' diagnosis of cancer was transformed into a potential revelatory moment in which hope might be instantaneously lost."[14] In 1912, Dr. Henry Prescott referred a 42-year-old breast cancer patient to Dr. Richard Cabot, chief of staff at Massachusetts General Hospital. Because the woman showed "signs of lung involvement" as well as loss of appetite and nausea, the disease was clearly advanced and "an operation seemed unwarranted." Asking Cabot for a second opinion, Prescott wrote that the patient had "been told that the tumor is benign."[15] Doctors who revealed cancer diagnoses often exaggerated the promise of the therapies administered and denied evidence of recurrences.[16]

Unlike many of his colleagues, Dr. Lawrence F. Flick adamantly refused to conceal tuberculosis diagnoses. A Philadelphia physician who himself suffered from the disease in the early 1880s, Flick not only treated large numbers of private patients but also played a leading role in the early-twentieth-century tuberculosis control movement.[17] "I make it an absolute rule," he wrote to a physician seeking an appointment for his wife, "to decline to see patients who are to be kept ignorant of their condition."[18] But if Flick revealed diagnoses, he carefully camouflaged prognoses, asserting that better obedience to his methods definitely would result in a cure. He dismissed some patients' setbacks as unimportant and blamed noncompliance for others, which he said would abate as soon as behavior changed. Only when death was hours or days away did Flick acknowledge the possibility of therapeutic failure.[19]

To be sure, sympathy and compassion could survive medicine's growing professionalization and concealment, especially when doctors practiced outside hospitals. As late as 1926, more than 70 percent of patients in rural areas of New York died in private dwellings.[20] Some of the most famous anecdotes about William Osler testify to the tender care he bestowed on patients dying at home. The mother of a girl who died in the influenza epidemic of 1918 later recalled the "delicacy and careful attention" he had displayed during his last visit.[21] Nevertheless, the thrust of medical education combined with increased secrecy often discouraged such sensitivity.

The training of nurses also emphasized the importance of maintaining emotional distance. Historian Barbara Melosh notes that although the supervisors of student nurses "tacitly acknowledged students' emotions in the face of death," they "insisted on self-control. One nurse's memoir described her lonely watch with a young dying patient and her overwhelming emotion as the end approached. She resolved, 'I must steady myself. It will never do to go to pieces like this at the last moment.' A nursing student who later confessed that she 'became totally unstrung' after a patient died also noted that she eventually learned to restrain her feelings of vulnerability."[22] Such struggles for control may have led nurses to further armor themselves against the suffering they witnessed.

Nurses learned to cultivate detachment from individual patients as well as from death. That task was especially important for the many nurses who worked in private duty. Tuberculosis nurses repeatedly complained about overly affectionate family members who undermined the therapeutic regime. One nurse reported to the supervising physician that the patient's wife "has annoyed me greatly for the last month because there is no sick room rule here. . . . You have supported me in saying Mr. M. must have rest and quiet. But frequently Mrs. M. insists on my leaving the room 'as she knows very well how to care for him.' " The nurse stressed her duty to "not yield to the authority of a patient's relatives or their kindness." [23]

And yet personal relationships inevitably developed. When journalist Charles Dwight Willard lay dying of tuberculosis at home in the early twentieth century, he initially described his nurse in terms of her professionalism, boasting to his family that Mrs. Delive was "one of the best trained nurses [he] ever saw or heard of." She had been one of the best students at Mt. Sinai Hospital in New York and had accumulated long experience since then.[24] As Willard's illness lingered, however, ties of friendship increasingly bound patient and nurse. He noted that Mrs. Delive was an attractive widow and teased her about her many beaus. He also occasionally worried that the demands of care overwhelmed her. She, in turn, visited Willard on many days for which she was not paid. "No one could be more faithful and devoted," he wrote after she had been in attendance for several months.[25]

Public health nurses similarly responded to patients in various ways. Interviewed by the Federal Writers Project in 1939, one woman recalled an incident at the beginning of her nursing career in 1914. Called to attend a young man living in what she described as a "shack," she was "startled to find the young man not only highly intelligent but extremely well educated as well." As he got worse, both she and the doctor "did everything humanly possible but in vain." She concluded: "That was the most harrowing case I ever undertook. I was only twenty-five myself at the time, and I guess I made a nurse's fatal mistake of caring. I wanted so desperately for the man to live."[26]

That nurse's thinly veiled contempt for the poor patients on her list is unsurprising. Middle-class women commonly recoiled from the impoverished clients they visited at home. A few, however, were deeply moved by the tragedies they could not prevent.[27] Writing to Lillian Wald, the founder of the Henry Street Visiting Nursing Service, Ada Beazley provided "a little account" of the tuberculosis death of an Italian woman in 1916. The husband was a longshoreman; the six children ranged in age from 9 months to 10 years.

> The pulmonary hemorrhage was practically continuous but devoted care was given by the husband. The patient could take nothing for several days before she died. Saturday Dec. 23rd, nurse arranged for Xmas gifts and dinner. There was never any evidence of aught besides bread, milk and coffee on the table. When the nurse arrived that morning the mother was dying. With her last conscious look, the sweet patient was grateful for having her mouth cleansed and a clean night dress and sheet. The fire subdued and silently weeping children were heart breaking (imagine crying quietly). Their attention was somewhat distracted by the parcels in the kitchen that arrived just as their poor father took the crucifix after dusting it off on his trousers and placed it above his wife's

> head. Handling a lighted candle to the nurse he held one himself and invested in mystery and majesty that soul passed away.

Beazley concluded that she "value[d] highly the privilege of serving this family."[28]

"Houses of Health"

As hospitals increasingly oriented themselves around science rather than social welfare and religion, autopsies gained new significance. One prominent Boston physician proclaimed: "A high percentage of autopsies on a physician's hospital service is an indication of a progressive attitude toward scientific medicine. The percentage of necropsies is the best simple index of the professional efficiency of the hospital."[29] Closer ties with medical schools, where anatomy had a dominant place in the curriculum, helped to elevate the importance of postmortem examinations. During the early twentieth century, the influential journal *Modern Hospital* was replete with articles urging administrators and physicians to obtain permissions for autopsies from family members and explaining how to do so.[30]

The new concern with the bodies of the dead, however, did not translate into greater attention to the needs of the dying. Although accounts indicate that some small religious hospitals continued to provide tender care at the end of life, most facilities emphasized the development of specialized services and out-patient clinics, the pioneering discoveries emanating from their refurbished research laboratories, their ability to attract eminent physicians, and their affiliations with prestigious medical schools.[31]

The need to woo a middle-class clientele gave hospitals an added incentive to present themselves as places for the production of health. Economic pressures forced administrators and trustees to adopt business models of financial management and rely more heavily on the fees of private patients. Hospitals thus designed services to attract the middle-income patients who previously had shunned those institutions. Exaggerating the extent to which nineteenth-century hospitals has been sites for dying, institutions now emphasized their ability to cure. In 1918, the prominent hospital architect Edward F. Stevens wrote: "The people are realizing that the hospital is built to benefit humanity and not to afford a place to die. 'All hope abandon ye who enter here' no longer is the appropriate inscription for the hospital age."[32] The publicity departments that hospitals inaugurated throughout the twenties continually reinforced that point. "The big idea," wrote S. S. Goldwater, director of New York's Mt. Sinai Hospital, "is to get the doctor and the hospital established in the popular mind in association with successful efforts for the preservation of health."[33] A 1921 *Modern Hospital* article urged hospital managers to "take a lesson from the clergyman who advertised his church as a 'Cheerful Church.' Publicity can make hospitals 'houses of health.'"[34] An article ten years later advised: "It is eminently worth the while of any superintendent to take the trouble to tell the local newspaper each morning that Mr. and Mrs. William Brown of Park Avenue are the proud parents of a daughter born at the hospital. He may offset the chill of the death notices and silence the comment that is frequently made that 'Nobody ever leaves the hospital except feet first,' by giving a list of the patients that have been discharged to their homes."[35]

Elaborate building and landscaping programs offered new opportunities for hiding the presence of death. One historian notes that architects employed "hotel and residential imagery as a mechanism disguising the hospitals' association with illness."[36] We can assume that the intent also was to render death completely invisible. Because the increased emphasis on autopsies meant that morgues grew in size and in number, concealment became especially urgent.[37] Edward Stevens stressed that because the death rate in tuberculosis wards "will naturally be greater than in any other part of the institution, the method of removing the body from the building so as to attract the least attention should be studied. If the morgue can be at some little distance from the wards, with an underground connection, much mental suffering will be avoided."[38] In a 1922 article on recent changes at Newton Hospital in Massachusetts, the author wrote: "In the early days the mortuary with all its depressing effect had been erected on the grounds adjacent to several of the wards." Fortunately, the contour of the land "lent itself most admirably to the correction of this defect. A new brick building was erected between the two low hills. The entrance to the mortuary and chapel was made on a lower level reached by a driveway behind one of the hills and outside of the main hospital grounds. On the higher level the driveway within the grounds leads to the ambulance house which is the second story of the mortuary building. Nothing but the ambulance house can be seen from the hospital but there is a direct connection through an underground corridor." The caption of the accompanying photograph read: "The trim utilization architecture of the ambulance house gives no hint to patients in the buildings across the green that it conceals a morgue."[39]

Changing internal architecture further helped patients avoid contact with death. In an attempt to attract a wealthier patient population, new hospitals replaced large wards with small wards, semiprivate rooms, and private rooms. As a result, patients less commonly witnessed the deaths of others. New facilities also typically included isolation rooms, where nurses could put ward patients in extremis. A physician explained why it was especially necessary to segregate patients who were dying of cancer. Such patients "sometimes are left in the midst of the living for hours and sometimes, in lingering cases, even for days. The psychic effect on other patients is distressing in the extreme and sometimes days pass before the ward is restored to its normal composure." That was "not as it should be." "As soon as the signs of dissolution become manifest the patients should be removed to a room set apart for the purpose. . . . Other occupants of the ward are thus spared the harrowing scenes which so often attend the death of cancer victims."[40] When separate rooms were unavailable, nurses pulled the curtains or placed screens around dying patients.

But isolation rooms, screens, and closed curtains also had disadvantages. An 1892 *New York Times* article explained why one New York hospital had decided to abandon the practice of taking dying patients out of the ward: "Patients soon learned the significance of a removal to the [isolation] room and when they saw a patient taken they knew that death was imminent. The knowledge was thought to be as depressing as the old practice of letting a patient die in the ward."[41] The meaning of closed curtains was equally clear. Lewis Thomas recalled his experience as a medical student in Boston City Hospital in the mid-1930s: "The white curtains around each bed were usually kept to one side, but when a death was about to occur the head nurse would make the rounds of the ward, moving fast, pulling each curtain across the foot of every bed. I remember the zinging sound of those curtains being yanked across on their metal rings. . . . The sound was the sound of dying, and all the other patients, the day's survivors, knew what it signified."[42]

Attempts to protect other patients also alarmed the moribund. In 1921 the son of a woman who died at St. Luke's Hospital complained that the "death screens" placed around her bed had terrified her.[43] Although Carlos Bulosan, a Filipino immigrant, survived his bout with tuberculosis, he came close to dying. In his semifictional autobiography, he described his long stay in Los Angeles County Hospital during the 1930s. When his condition sharply deteriorated, he was transferred to a small room where three other men were "waiting to die." He watched them as they lay "gasping for the last bit of air."[44]

A major justification for the use of private rooms was that they avoided such problems. A 1912 article in *Modern Hospital* focused on the benefits to patients who were likely to get better: "The patient in the hospital is always under nervous strain. He is all atingle with excitement when the doctors begin crowding around one patient, when the nurses hurriedly and ominously surround a bed with the white screens. He knows that danger is threatening; he hears the shriek of the patient in agony and then the death rattle. Surely these are not conducive to recovery. In the individual room all these annoyances are obviated."[45]

Reducing Mortality Rates

Private hospitals minimized their association with death primarily by concentrating on conditions from which patients were most likely to recover. As in the past, the great majority of hospitals showcased their low mortality rates (still typically less than 10 percent). "The sure test of efficient hospital management," wrote the medical director of Philadelphia's Chestnut Hill Hospital, "is found in the death rate."[46] Most administrators must have agreed because they presented their rates with little explanation. A few elite institutions, however, argued that high mortality rates demonstrated excellence because they implied that the physicians had accepted difficult cases. "What is success in hospital work?" asked New York's General Memorial Hospital in 1905. Although the first answer might be "a diminished death rate," this is not necessarily true: "Minor cases mean a small death rate. But the successful treatment of major cases may much better illustrate the advantage of hospital treatment, even although figures point to an enlarged number of cases of those who have been discharged as cured."[47] Other institutions justified relatively high rates by pointing to the impoverished living conditions of their patients and the poor care they previously had received.[48] And some facilities argued that other hospitals kept their rates artificially low. "There are institutions in this city where patients are not allowed to die on the premises, if it can be avoided," wrote an attending physician at Babies' Hospital in New York in 1900. "To accomplish this requires a great deal of chicanery. The cases are first culled with the greatest possible care, and if then they do not thrive they are promptly sent to other institutions."[49]

There was much substance to that complaint. Economic considerations helped encourage hospitals to retain traditional methods for reducing mortality rates, denying admission to many people with terminal conditions and discharging others when death seemed imminent. The advent of the ambulance had complicated the problem of limiting the number of patients dying from acute conditions.[50] In June 1869, Bellevue Hospital established the first civilian ambulance service. Both private and public facilities throughout the country soon emulated that example.[51] A high proportion of the patients brought by ambulance were those whom private hospitals did not want—not just severely injured or gravely ill but also very likely to need free care. As one contemporary observer wrote, the "ambulance subject" was "usually a person in poor circumstances. One rarely sees a well-dressed occupant being carried to the hospital by ambulance."[52]

In the mid-1880s, the *New York Times* began to report that private institutions were shifting patients who arrived by ambulance to Bellevue. For example, when

8-year-old Julia Bictor was run over by a coal cart one December morning, an ambulance delivered her to Chambers-Street Hospital. But at midnight, "the little sufferer was placed in an ambulance . . . and driven over three miles to Bellevue Hospital. . . . Five minutes after reaching the hospital in Twenty-sixth street the child expired." The night captain at Bellevue stated that although he had "seen many hard cases," "this one was enough to touch the heart of a stone. . . . The poor little thing had no covering but a muslin bandage around its arm and waist, and sending a patient out in the night air in that condition certainly did not prove of benefit to it."[53] In 1891 the *Times* ran a story entitled "Death Caused by Removal." This time German Hospital "hurried a dying man from their institution to Bellevue, where he died seventeen hours after admission." Although German Hospital authorities pointed out that they had "loaned" the ambulance to the man's father "as an act of charity," Bellevue Hospital physicians called the patient's "dumping" "an act of inhumanity."[54]

In some cases, private hospitals instructed their ambulances to take critically ill patients directly to Bellevue regardless of the distance. When the wife of an ice-wagon driver suffered serious burns on her back, face, and limbs in January 1897, the police summoned a Roosevelt Hospital ambulance, which then rushed her to Bellevue, although Roosevelt "could have been reached more quickly and over a smooth asphalt pavement." There was a "grave possibility," the *Times* concluded, "that the woman might die as a result of her injuries."[55]

When patients arriving in private hospital ambulances died either en route or soon after entry to Bellevue, coroners investigated and in some cases issued censures. In 1902 Commissioner of Public Charities Homer Folks ordered that he be sent a full report from the superintendent of any hospital in which a patient transferred from another facility died within three days after admission. The following year, the Department of Public Charities announced: "As a result the number of such transfers has markedly diminished, and there are now practically no transfers of this character except under circumstances which make the transfer practically unavoidable."[56]

But the department spoke too soon. In March 1906, according to the *Times*, an ambulance transported a dressmaker with heart disease to New York Hospital. "There she became rapidly worse, but in spite of her condition she was put back into the ambulance and orders were given to take her to Bellevue. She expired on the way."[57] That case so "aroused" the chairman of the Manhattan Board of Coroners, Julius Harburger, that he pressed a state senator to introduce a bill providing that any hospital official who transferred an ill patient to another institution would either be fined $5,000 or imprisoned for five years.[58] After that bill met defeat,

Harburger proposed a city ordinance to impose a penalty of $100 in every case where a hospital superintendent either refused admission to a sick patient brought in an ambulance or ordered the transfer of a critically ill patient. When private hospitals protested, Harburger exploded in rage, noting that their policies affected only patients who could not pay. Private hospitals were murdering the poor, he shouted. "Open your books, and show me one man of wealth who has ever been transferred."[59]

The city ordinance eventually passed, but it was only partially effective.[60] In March 1914, the Hospital Investigating Committee of the Board of Estimate reported: "That eighteen patients brought to Bellevue in private ambulances (in the three-month period under consideration) died on the day of arrival indicates a marked tendency on the part of private hospitals to carry dying patients to Bellevue rather than their own hospitals." Furthermore, the report noted, private hospitals in Brooklyn were "even more prone" than those in Manhattan "to transfer patients on the point of death" to King's County Hospital, the borough's public facility.[61]

Chronic Diseases

Admission and discharge policies directed toward people with advanced chronic conditions provoked less outrage than the transfer of patients gravely ill with acute diseases but may have been even more significant. As the rates of many acute diseases declined between 1870 and 1940, the proportion of deaths caused by chronic diseases steadily rose.[62] Nevertheless, the average length of stay decreased, and sufferers of long-term conditions continued to represent a small fraction of the hospital clientele.[63] By 1900, most general hospitals admitted people with cancer, but only when they needed curative treatment.[64] Those who failed to improve were discharged as soon as possible. According to a 1929 New York City study, "The cancer mortality in the general hospitals is small, as comparatively few patients are held until death."[65] Like the critically ill acute-care patients that ambulances delivered, many with incurable diseases were quickly transferred to public facilities. Interviewed by the Federal Writers' Project in 1939, an intern remarked: "City Hospitals get all the 'crocks' . . . that is, patients who are really what we call chronic, medically typified as chronic, heart disease cases and others for which cures are unknown, and the only treatment, therefore is palliative . . . for instance, cancer patients who are pretty far gone."[66]

The argument private hospital administrators most frequently invoked to explain attempts to weed out patients with such chronic diseases was efficiency. During the six months one such person might occupy a bed, the hospital could care for as many as fifteen people with acute diagnoses.[67] Observers noted, however, that other explanations were equally compelling.[68] One was that attention focused even more exclusively than before on treatments for acute illnesses. Assuming that their graduates rarely would interact with chronic disease sufferers, medical and nursing schools wanted to affiliate only with hospitals filled with patients whose problems could be quickly as well as successfully resolved.[69] Hospitals serving large numbers of patients with long-term problems often had difficulty attracting visiting physicians.[70] "Medical life in the general hospital must be exciting and dramatic," wrote the director of one of the few private chronic care hospitals. "The stimulus of quick results—the kind of results that are so pleasant and so rewarding when reported at meetings of medical societies—is a kind of compensation for voluntary service by visiting staff in hospitals."[71] A Los Angeles physician explained that his typical colleague was happy to volunteer in the acute medical or surgical department of the county hospital because "he gets an immense amount of valuable education for himself." Patients with diseases such as tuberculosis, by contrast, were "like the poor; you have them with you always. You can take a man and treat him for an ordinary disease, have him there ten days or two weeks and get rid of him. We don't mind spending a little money to do that, but when you take a case of tuberculosis you have him there months and months, possibly for years."[72]

And not only were chronic care patients "like the poor"; they also often were the poor. As the following chapter will demonstrate, prolonged disease frequently led to impoverishment. The admission of large numbers of people with chronic conditions to hospitals thus would have impeded the goal of attracting a predominantly middle-class clientele.

Dying in a Private Hospital

Private hospitals could never completely rid themselves of death. During the 1918 flu epidemic, dying patients filled the beds of all hospitals, regardless of their admission criteria. "It was not unusual to pass two or three of your patients being carried out the back door as you were going up to make your midnight visit," recalled a medical student at Massachusetts General Hospital.[73] A nursing student at Mt. Sinai in New York had an equally grim experience: "When a patient ceased, there was another one there to take his place before we could change the bed. We tried to care for the urgent ones first, but it was hard to tell which ones they were."[74]

But even during normal times, death continually visited hospitals. Although administrators tried to emphasize conditions most likely to produce success, expectations often outran reality. (Indeed, despite widespread publicity touting the safety of hospital births, some evidence suggests that hospital deliveries were more dangerous for women throughout the 1920s.)[75] Moreover, as New York coroner Harburger pointed out, hospitals unloaded only dying patients who were poor on public facilities; administrators might encourage wealthy patients to die at home, but they rarely were forced to leave. And private hospitals could transfer only a small fraction of poor patients in extremis; most died in the facilities they originally entered.

The reticence surrounding hospital deaths makes it difficult to glimpse the kind of care dying patients received. Hospital regulations often specified the responsibilities of doctors and nurses after death occurred, but most were silent about the obligations of medical staff toward patients approaching that event. Trustee minutes also included few discussions of that topic. And while there were many *Modern Hospital* articles on the need to obtain autopsies, there were virtually none on the needs of patients facing death.[76]

It is unclear to what degree dying patients received solicitude and compassion. Some evidence suggests that they got little. As we have seen, professional training emphasized the importance of maintaining emotional distance from patients, not developing intimate attachments with them. Shorter lengths of stay may have increased the sense of detachment. A nurse at New York's Roosevelt Hospital in the early twentieth century later wrote: "To us each [patient] was an individual whose likes and dislikes, characteristics, problems, joys, and sorrows we quickly came to know. . . . We rejoiced when they recovered and went home. . . . When they died in spite of all we could do, there remained an unexpressed heartache."[77] When death came at the end of a brief stay, however, nurses were less likely to feel a sense of loss.

In some cases, family members could compensate for staff impersonality. Hospitals continued to post "danger lists," containing the names of all critically ill patients. Regulations directed staff to notify relatives as quickly as possible and allow them to stay by the bedside outside regular visiting hours. "No effort should be spared to get word to the friends of a patient when it is seen that the fight for his life is hopeless," wrote S. S. Goldwater, director of Mt. Sinai. Furthermore, he stressed, "It is not enough to notify the relatives of such a patient that they may visit the hospital at will and remain as long as they like; their comfort must be looked after while they are there. I know of one hospital that has made many friends for itself by declining to accept fees for furnishing food or shelter to the friends of dangerously sick patients."[78] Other administrators pointed out that family members who received special accommodations were more likely to agree to autopsies and less likely to charge that negligence had occurred.[79] It would be impossible to determine what proportion of dying individuals had families in attendance: some case files indicated whether relatives and friends were present when breathing ceased, but many did not.[80]

We do know, however, that the presence of family members often depended on the location of patients and their ability to pay. When death occurred on the wards rather than in separate rooms, relatives disturbed other patients. Families also interfered with medical routines. When one grieving mother complained about the inadequate treatment her dying son had received at New York's Flower Hospital in 1911, an investigation concluded:

> It is only natural that his poor mother . . . should feel that something should be done every moment, and that the Nurse and Doctors should constantly hover over his cot. As a matter of fact, this distracted lady continually fretted and fussed and interfered with the directions of the attendants. If the Nurse left the bed-side for a moment, on her return she would find the mother lifting the head and shoulders of the boy, and even bolstering up the body of a patient whose very life depended upon absolute quiet. If any mistake at all was made in this case, it was in permitting the friends and relatives so much freedom in attendance.[81]

It is likely that family members most frequently maintained vigils around beds in private rooms, which tended to have liberal visiting hours even when patients were not gravely ill. In 1913, for example, New York's General Memorial Hospital for the Treatment of Cancer and Allied Diseases permitted visitors between 8:30 a.m. to 9:00 p.m. in private rooms but only between 1:30 to 3:00 p.m. in wards.[82] Some hospitals allowed relatives and friends of private patients to rent adjacent rooms to sleep at night.[83]

Relatives of patients dying in wards occasionally voiced serious complaints. Some stated that administrators had failed to notify them in time. A few received misinformation. According to the February 9, 1900, *New York Times*, an ambulance "knocked down and fatally injured" Margaret Flood, a 52-year-old widow. "Her skull was crushed in two places, the horse having stepped on the left side of her head, and there was a fracture of the pelvis and other internal injuries." The hospital sent a note to her daughters that she was "somewhat dazed" and had a "slight scalp wound" but that her condition was otherwise good. Even after the daughters arrived at the hospital, they were told that Mrs. Flood's condition was not serious.[84]

Criticisms also arose about the lack of consideration accorded to families during the hours preceding and immediately following death. One woman whose father had died in a Massachusetts General Hospital ward in 1911 wrote that the relatives of critically ill patients should receive "the attention and courtesy due them." She had been forced to hold her brief conversation with a doctor in a "common entrance way."[85] The employer of a man who died in New York's Flower Hospital told the *Times*:

> The so-called waiting room opens directly off the main hall of the hospital. People were walking through the room continually off the main hall of the hospital. . . . It was in this room that [the patient's mother] had to wait after his death for some two hours—in a place where strangers were passing in and out continually, laughter and boisterous talk was going on, and where, just before she left, a crowd of medical students came in. There is apparently no accommodation in Flower Hospital for any privacy to be given to any friends or relatives of patients who may be so unfortunate as to be in the so-called charity ward.[86]

Care in Public Institutions

Although more than half of all hospital beds were in private facilities in 1929,[87] most hospital deaths probably took place in public institutions. One reason is that some public hospitals were so overcrowded that they contained far more patients than beds. Mattresses commonly lined the floors in halls and wards. "There are at present 623 beds and cribs in the buildings," reported New Orleans's Charity Hospital in 1883. "This number is 100 more than the cubic capacity of the wards will properly allow. Yet, in most wards, all the beds are nearly always occupied, and patients sleeping on the floor. This overcrowding is a complaint of long standing."[88] The problem was even worse in the mid-1930s, when the hospital routinely placed two people (and occasionally even three) in a single bed. As a result, the facility listed 1,814 beds but an average daily patient census of 2,781.[89] In addition, as the administrators of public hospitals frequently complained, their mortality rates were increased by transfers from private facilities.

The establishment of affiliations with both medical and nursing schools and the purchase of new technological equipment helped to upgrade public hospitals as well as private ones.[90] But because meager budgets kept staffing levels low, dying patients in all public hospitals, and especially those with large chronic populations, received little care.[91] In 1914 the director of the California State Bureau of Tuberculosis found "a decided shortage in the nursing force" at Los Angeles County Hospital, "the management being unable to obtain trained nurses, so that in the ward for the most advanced cases there was but one trained nurse who served eight hours a day. The remainder of the work was necessarily done by untrained orderlies."[92] A 1924 New York study reported that municipal hospitals gave the "lowest amount of nursing per patient."[93] Bestowing special attention on the dying must have been extremely difficult.

Pervasive views of public hospital patients may have made it easy for staff to disregard their needs. In 1908, the member of the medical board of a large New York

City municipal tuberculosis infirmary, housing many dying patients, stated that although some patients were "hard-working people," others were "of the lowest and most dissipated, the flotsam and jetsam of the social life of a great city."[94] An intern at Philadelphia's almshouse hospital wrote about one of his patients close to death, "The unfortunate being certainly has suffered horribly, and would arouse every one's sympathy did he not also call up feelings of utter disgust for his character."[95]

Patients dying in public hospitals also had little privacy. A 1914 report on tuberculosis services in Los Angeles County Hospital commented: "The small number of private rooms for extreme cases is a defect which should be met. The wards for advanced cases should have at least as many private rooms adjoining as there are likely to be daily deaths. Since as many as six cases have died in one day on this ward, it is evident that the single room is quite insufficient."[96] Seven years later the director of the California State Bureau of Tuberculosis complained, "The dead have to be carried out in full view of all the other patients, and the overcrowding could not be any worse if one-half of Los Angeles had been stricken with a plague."[97] A doctor at New Orleans' Charity Hospital in 1937 recalled: "Imagine Mardi Gras with thirty-seven beds in a sixteen bed surgical ward. In trying to make rounds at night, I literally stepped over bodies of the sick lying in the aisles. It was not unusual for one of two patients in a bed to go tell the nurse, 'My bed mate has just died.'"[98]

Families were less likely to intervene in public facilities than in private ones. Although the regulations of public hospitals, like those of private institutions, directed staff to notify relatives when patients were in extremis and allow special visits,[99] the indigent clients of the New York Charity Organization Society frequently complained about the difficulty of obtaining prompt and accurate information.[100] In 1914, the New York City commissioner of public charities blamed "callous" nurses for the delay in getting word to a man's family that his condition was critical.[101] (He said nothing about the conditions of the nurses' employment, for which he was responsible and which may have explained some of the behavior he condemned.)The paucity of clerks at public hospitals must have caused other communication problems.[102] The families of patients who had recently been transferred from either a private hospital or another public one were especially likely to miss critical information. Because Polyclinic Hospital misspelled the name of a man when it transferred him to Bellevue Hospital, officials could not tell his wife he was dying. In his delirium, the patient could only utter the word "wife." He died alone, three hours after arriving at Bellevue.[103]

But even when hospitals alerted relatives in time, child care and work responsibilities often precluded trips to the bedside. Moreover, nurses sometimes registered their distaste for poor family members. In 1904, the New York City Charity

Organization Society asked a hospital to allow a woman to visit her dying son every day. The superintendent replied the mother could come but only if she could "remain quiet and calm." That mother must have complied, because the case report indicated that she stayed with the boy "both day and night" before he died.[104] Relatives who did not behave in a way the staff deemed appropriate may have been denied such an opportunity.

Challenging Medicine's New Direction

Although the horrendous conditions in public facilities typically escaped attention, a few medical personnel expressed concern about the growing impersonality of the health care system in general. Emphasizing the loss of "the sympathetic relation which formerly existed between doctors and their patients," a past president of the American Medical Women's Association urged her colleagues to avoid "narrowing and concentrating their vision upon the purely physical to the exclusion of the psychic and human."[105] An equally prominent critic was the chief of staff at Massachusetts General Hospital, Richard Cabot, who hired Ida M. Cannon, a recent social work graduate, in 1905. Ten years later, she established the first hospital social work department there.[106] Although Cannon's early writings contained occasional comments about "a hostile administration" and "the antagonism of doctors," Cabot offered unqualified support.[107] As other institutions emulated Massachusetts General's example, hospital social service departments gradually spread throughout the country. Like the institutions in which they struggled to gain a foothold, early-twentieth-century medical social workers rarely mentioned death and dying.[108] We will see, however, that one of their major tasks was to facilitate the discharge of patients deemed incurable by making other arrangements for their care.

Hospital chaplaincy programs represented another of Cabot's concerns. In 1925 he published *A Plea for a Clinical Year in the Course of Theological Study*, which helped to inspire the development of clinical pastoral education.[109] Encouraging theological students to spend a year in hospitals as part of their training, he argued, not only would enable the students to learn from those institutions but also would help to broaden their focus. The theological student would "look after the minds, the emotions, the wills, the souls of the inmates as the doctors and nurses now care for their physical welfare. Against this doctors and nurses will certainly insist that they are already caring for the whole patient, body and soul. But just as certainly, they are not doing so. Their attention is too strongly concentrated on the excessively difficult and delicate tasks of diagnosis and treatment. There is not enough attention left to go around."[110]

In addition, Cabot collaborated closely with the Reverend Russell L. Dicks, who first arrived at Massachusetts General as a patient. As Dicks gradually improved, he began to visit other patients. According to one widely circulated story, Cabot remarked: "Here is a man who writes down the prayers he had with a man who is dying! We'd better ask him to stay . . . we might learn something."[111] Cabot appointed Dicks chaplain at the hospital in 1933, and together they wrote *The Art of Ministering to the Sick*, which appeared three years later. In a chapter entitled "The Dying," Dicks noted that his recent article, "Religion's Contribution to the Recovery of Health," had included one short sentence about death. Hoping to correct that imbalance, Dicks described his many interactions with patients facing death and recorded the prayers he recited to them.[112]

The physical pain of dying patients also occasionally received renewed attention. After its 1887 publication, *Euthanasia; or, Medical Treatment in Aid of an Easy Death*, by British physician William Munk, attracted a wide readership on both sides of the Atlantic.[113] Like most of his contemporaries, Munk defined euthanasia as alleviating the suffering of death rather than accelerating it.[114] Medical intervention need not be extensive, he argued, because most dying people experienced no pain. "The process of dying and the very act of death," were "but rarely and exceptionally attended by those severe bodily sufferings, which in popular belief are all but inseparable from it."[115] When the eminent physician William Osler inaugurated his 1900 "study of the act of dying," one of his central goals was to test that conclusion.[116] As we saw in the introduction, Osler asked nurses at the Johns Hopkins University Hospital to record the experiences of patients during the last moments of life.

Osler appears to have gradually lost interest in his study. The deaths of fewer and fewer patients were documented each year, suggesting he did not encourage the nurses to continue to complete the cards he initially distributed.[117] Although a prodigious writer, he also failed to publish the study results, presenting just a brief summary of his findings in a 1904 Harvard lecture.[118] The 1929 essay "The Care of the Dying," by Richard Cabot's close friend and associate Alfred Worcester, was thus the first major medical publication on that subject since Munk's book more than forty years earlier. Like Munk, Worcester emphasized the peacefulness and painlessness of the dying and offered advice about relieving any physical discomfort they encountered.[119]

If medical social workers, hospital chaplains, and a few physicians sought to smooth the dying process, they could not fundamentally alter the reductionism of early-twentieth-century hospital care and its preoccupation with patients most likely to recover. For virtually all patients in public hospitals and the great majority of those in private hospital wards, dying remained an extremely lonely and dehumanizing experience.

Public Health Campaigns

Although public health never acquired medicine's power and prestige, it had an even more ambitious goal—to eliminate disease entirely by focusing on prevention. Such a mission demanded a transformation in popular attitudes about death. Public health workers relentlessly argued that mortality was not God's will, but rather a "sacrifice," the result of voluntary human action. In a discussion of the many deaths from diphtheria and whooping cough in 1915, a visiting nurse wrote that "hundreds of lives have been sacrificed because someone neglected quarantine rules."[120] After Schenectady, New York, officials rescinded a regulation to protect the milk supply in 1909, the Bureau of Health claimed that "46 babies had been sacrificed needlessly."[121] The refusal of an American Indian family to hospitalize a dying infant was, according to a public health nurse, "a deliberate sacrifice to false beliefs."[122]

Widespread campaigns at the turn of the twentieth century employed both hyperbolic language and graphic imagery to convince the public that death was avoidable and thus unacceptable. "Don't Kill Your Baby," implored a Chicago poster translated into Polish, Yiddish, Bohemian, German, Lithuanian, and Italian and plastered throughout the city.[123] Handouts distributed by the tuberculosis campaign, beseeching women to eradicate germs through household cleanliness, routinely linked "dirt, death, and disease."[124] And poster after poster depicted death as monsters, fiends, or devils, all familiar to a wide public as a result of the Gothic literature that was enjoying renewed popularity.[125] Where the literature implied that people could only shrink in horror before those figures, however, the posters suggested that public health regimes allowed everyone to resist and ultimately prevail.

The infant mortality movement exemplified efforts to alter public consciousness about death. Although that movement began in 1850, it gained extraordinary strength between 1900 and 1920. Simultaneously, the focus began to shift from milk purification to instruction of mothers.[126] Advocates of both orientations argued that infant deaths were not expressions of God's will but rather "needless," "preventable," and "avoidable." A few reformers used stronger terms. The adulterated milk that caused infant diarrhea and death had to be considered "murder" and "infanticide," according to the *Chicago Tribune*.[127] Julia C. Lathrop, the chief of the U.S. Children's Bureau, railed against the pervasive view that some early deaths were predestined and could not be prevented by human agency.[128] She told the 1919 International Congress of Working Women, "Science refuses to accept the old fatalistic cry 'The Lord gave, and the Lord hath taken away, blessed be the name of the Lord.' "[129] In an address to the American Child Hygiene Association the same year, she urged her audience to arouse the general public, which was "still in the dark limbo of fatalism."[130]

Reformers occasionally acknowledged how rapidly their own attitudes had changed. "We calmly accepted the annual harvest of [infant] deaths as if it were as inevitable as the weather; as if indeed a part of the weather," wrote Chicago health officials in 1910. " 'Hot weather, babies die,' was our unconscious thought."[131] Chapter 1 demonstrated that nineteenth-century clergy had praised the religious convictions that enabled people to reconcile themselves to the inevitable. But acceptance was no longer a virtue. "So long as mothers did not know that children need not die," wrote the social activist Florence Kelley, "we strove for resignation, not intelligence. A generation ago we could only vainly mourn. To-day we know that every dying child accuses the community."[132]

Reformers were especially quick to attribute fatalism and resignation to poor women. Josephine Baker, chief of the New York City Bureau of Child Hygiene, later complained that immigrant mothers were "horribly fatalistic" about infant death. "Babies always died in summer and there was no point in trying to do anything about it. . . . I might as well have been trying to tell them to keep it from raining."[133] The Children's Bureau reported that Lithuanian immigrant mothers in Connecticut assumed that some children were "ordained to die" in infancy.[134] According to the Social Service Bureau of Bellevue Hospital in New York, the mother of six-month-old Harvey had "given up." "Marasmus, malnutrition, and acute gastro-enteritis combined" had made him "a mere handful of bones and loose, wrinkled skin." Although "the mother cried over the baby," she "made no efforts to carry out the physician's orders." She took care of the baby when a nurse came to the house but "thought it was no use—then again it was so hot—anyway, Harvey would die no matter what was done"[135] A public health nurse in Arizona excoriated Gregoria, a 22-year-old Mexican mother: "Five babies had been born to her in the seven years of her married life, but the good God had taken three of them. They had lived a few miserable months, and then died. Gregoria sat huddled over her door-step and thought that nothing was of any use. The baby was always sick and some day the good God would take him as He had the others."[136] No longer an important spiritual goal, acceptance of God's will was now portrayed as a sign of dangerous passivity, inadequate willpower, and failure to appreciate the benefits of modern medicine.

Yet evidence of the enormous amount of care working-class mothers rendered to sick infants and children under extremely difficult circumstances helps to dispel the belief that they were invariably fatalistic and apathetic.[137] Moreover, public health workers often interpreted opposition to their programs as lack of concern for children's health even when such opposition made sense from the client's perspective.[138] Material conditions often prevented compliance with public health advice; clinic visits meant the loss of wages and sometimes jobs; and institutional placement devastated families emotionally and economically. But it also is possible that the reformers who chastised poor mothers for fatalism were trying to distance themselves from the older generation of women in their own families. Although few women advocates had children themselves, many remembered mothers, aunts, and grandmothers whose offspring had died at young ages.[139] "I was the third of eight children, healthy boys and girls," Florence Kelley recalled. "After the death in 1859, of my elder sister Elizabeth, aged two years, entries in the family Bible followed with pitiful frequency . . . : Marian in 1863, aged eleven months; Josephine in 1865, aged seven months; Caroline in 1869, aged four months; and Anna in 1871, aged six years."[140] In response to so many losses, Florence's mother developed "a settled, gentle melancholy which she could only partly disguise." According to Kelley's biographer, the mother's sadness threatened to engulf her one remaining daughter, who embraced political and social action partly as an antidote.[141] Any resignation she perceived in other mothers may have seemed dangerously akin to the depression and passivity she had seen in her own.

The following chapter examines the impact of public health campaigns geared toward the three chronic conditions that claimed the most lives: tuberculosis, heart disease, and cancer. By encouraging early detection and treatment, those drives may have helped to prevent many deaths. But just as the infant mortality movement increased the blame directed toward mothers of dying children even while averting many deaths, so the campaigns organized around specific diseases heightened the stigma surrounding those who had incurable conditions. Dying people who were poor and members of ethnic and racial minorities drew especially fierce condemnation.

CHAPTER 4

Institutionalizing the Incurable

"Cancer is getting to be a common disease," observed Minnie Patterson in a 1912 letter from her new home in Alaska to her Illinois family. "It seems there are always one or more afflicted with it in this town or community." A 90-year-old man buried the previous week had died from "the effects of cancer on his face." A middle-aged woman recently had had "both breasts cut off, in two different operations." Although she still was "able to be around," Minnie feared the worst: the disease would "break out some other place before long."[1]

Minnie's concern about the growing risk of dying from cancer soon received statistical confirmation. Drawing on data collected between 1911 and 1922, researchers at the Metropolitan Life Insurance Company reported in 1925 that the cancer death rate had significantly increased.[2] "We are now confronted with a new situation," Louis I. Dublin, the company's chief statistician, wrote two years later. With "the reduction in the mortality from other conditions . . . [and] with every improvement in the condition of life in the early ages, more and more people will approach the later period [of life] when the population is exposed to the cancer menace."[3] By 1940, cancer caused 165,000 deaths annually in the United States.[4]

Despite his initial focus on cancer, Dublin was well aware that other chronic conditions added even more significantly to the death toll. Heart disease was the leading killer, responsible for 385,000 deaths in 1940; hypertension and arteriosclerosis accounted for another 144,000.[5] The only major chronic affliction to decline in importance was tuberculosis. When Met Life issued its 1925 report, the disease still surpassed cancer as a cause of mortality, but the TB death rate already had begun to wane.[6]

Redefining the Incurable

Although the new enthusiasm about medical prowess focused primarily on acute illnesses, it did not entirely bypass either tuberculosis or the two long-term, noninfectious diseases capturing increased attention. Physicians and public health officials declared that all three afflictions were curable if detected and treated early enough. Permanent recovery from tuberculosis was extremely difficult before the 1946 introduction of streptomycin, but Robert Koch's discovery of the tubercle bacillus in 1882 had greatly increased medical and public health confidence. The New York Department of Health declared consumption in 1904 to be preventable and curable. "In a majority of cases," the department stressed, "it is not a fatal disease."[7] Patients must, however, get help in time. A prominent tuberculosis specialist wrote that the sufferer who postponed medical care "knocks at the doctor's door, and the undertaker answers."[8]

Turn-of-the century surgical advances greatly inflated the confidence of cancer specialists. Established in 1913, the American Society for the Control of Cancer (ASCC, later the American Cancer Society) sought to counter the pessimism Minnie Patterson had expressed. Although Minnie had assumed that her neighbor's double mastectomy would be powerless against the onslaught of the disease, the new organization began to tout the ability of early detection to prevent death. By the 1930s, when approximately four-fifths of cancer sufferers still died of the disease, a massive health education campaign promised that cancer would not kill anyone who remained vigilant and reported all suspicious signs to doctors.[9] An article in the *American Journal of Nursing* encouraged nurses to convey "the message of hope and cheer that *early cancer is curable* and only *neglected cancer is incurable*."[10] The title of a widely distributed ASCC film proclaimed *Choose to Live*, implying that the outcome of disease was entirely under individual control.[11]

The development of the electrocardiograph infused optimism into the field of cardiology. Although effective treatment lagged far behind, the American Heart Association, established in 1922, advised everyone to get annual exams to detect heart disease in its early stages. John Wycoff, one of the founders of the new association, noted that the organization's aim was to combat the widespread view that "arteriosclerosis, like a fate, must attack us all."[12]

A major result of the early detection drives was the separation of patients into two distinct groups—those with incipient disease, who were deemed curable, and those with advanced or what we now call "terminal" conditions, who remained incurable. Members of the second group increasingly were held responsible for their own misfortunes. Doctors who previously might have viewed advancing disease as either inevitable or the result of their own incompetence instead began to blame patients for having failed to recognize the first symptoms and take corrective action.[13]

Other factors intensified the denigration of incurable patients. Traits commonly associated with disadvantaged people were believed to make them especially prone to delay. Cardiologist Louis Bishop observed that patients in public clinics and hospitals tended to ignore the "minor discomforts" of incipient heart disease and "were in a great measure fatalistic in their outlook upon life." By contrast, the wealthier patients he saw in private practice "did not disregard any personal discomfort" and faithfully adhered to "any reasonable plan of treatment or change in environment."[14] The author of a 1927 article wrote that "among the more ignorant women of the laboring classes, cancer of the womb is almost never diagnosed in time to cure."[15] Race as well as class distinguished patients who postponed diagnosis and care. A Virginia physician explained why only "a very small percentage of Negro consumptives ever recover": "Their childishness . . . keeps them in the first place from being willing to acknowledge they have consumption and later on from being willing to persevere in the treatment."[16] A University of Southern California social work student quoted with approval an expert who declared that the Mexican resembled the Negro, a "happy-go-lucky person who is not as profoundly disturbed by the early stages of tuberculosis infection as the average white."[17]

Eugenic beliefs intersected with cultural ones. Viewing cancer as a disease of "civilization," most experts associated the disease with whites. African Americans were assumed to be too "primitive" to live long enough to contract the malady in large numbers.[18] But the notion of primitiveness worked very differently in the case of TB. Although Koch's breakthrough had punctured the belief that the disease could be directly inherited, many experts accepted the conclusion of the British eugenicist Karl Pearson that people inherited a predisposition to the disease. A related argument was that both African Americans and American Indians experienced extremely high rates of TB because they were primitive people who lacked prior exposure to the disease and thus never had developed immunity. The same traits that increased vulnerability to disease were believed to diminish the ability to fight it. In a discussion of tuberculosis in the Southwest, Dr. Ernest A. Sweet, a former surgeon in the U.S. Public Health Service, noted that the disease advanced especially rapidly among Mexican immigrants: "Recoveries are exceedingly rare, most physicians confessing never to have seen one, and the course is almost invariably progressively downward. A person will be about his work apparently well, suffer from a hemorrhage, and in four months will be dead." By contrast, Sweet insisted, Mexicans who were "less contaminated by Indian blood" exhibited "far more resistance to the disease."[19]

The growing embrace of the germ theory also transformed conceptions of people with tuberculosis.[20] Just as Americans began to learn that cancer was not contagious, health officials launched massive education campaigns warning that TB sufferers represented a dire threat to others. "Advanced cases," which produced the most "poisonous" sputum, were assumed to pose special perils. And health officials relentlessly asserted that because patients from disadvantaged groups were too "careless" to follow health advice, they were most likely to spread disease.[21]

The high premium placed on economic efficiency and independence in American society further heightened the hostility toward patients with all incurable conditions. Partly because charitable organizations quickly dropped able-bodied adults from caseloads, families with chronically sick breadwinners represented a very high proportion of clients.[22] Many also seemed to drain government coffers by spending long periods in public hospitals and sanatoriums. To be sure, some commentators pointed out that disease and disability were often the main cause of poverty. The "economic injury" wrought by tuberculosis was a particular source of concern to early-twentieth-century charity workers. In 1916 the New York Charity Organization Society published a study of New York families on relief at tuberculosis clinics. In twenty-one families, the man was living but suffering from sickness. Only four of these men continued to hold their jobs; seven were unemployed, and ten had irregular jobs with lower pay.[23] Twenty years later, the social reformer Grace Abbott quoted a government report which concluded that "one-third to one-half of all dependency can be traced to the economic effects of illness."[24] Nevertheless, patients with advanced disease encountered special opprobrium because they could not regain "workforce capacity," an increasingly critical measure of human worth.

Finally, the late stages of both tuberculosis and cancer produced conspicuous symptoms at a time of growing concern about the presence of visibly sick and disabled people in public spaces. Dr. Sweet declared, "The far advanced consumptives . . . are objectionable, those who present no appearance of invalidism being in an entirely different category from those who exhibit every evidence of the ravages of disease."[25] In the spring of 1910, Charles Dwight Willard, a journalist with tuberculosis, boarded a Los Angeles streetcar and saw a woman he later would describe as severely deformed; her spine was "twisted" and her head jiggled constantly. Just as he was thinking "what a hideous affliction" she had, he realized he appeared equally offensive to her.[26] His symptoms included not only extreme emaciation but also frequent, uncontrollable coughing.[27] The previous year, Pasadena residents had signed petitions urging the Los Angeles County Board of Supervisors to close a health camp for indigent patients, arguing not only that they represented a danger to the rest of the community but also that "the coughing of the patients is constant night and day, and . . . we do not think it fair or right to be compelled to hear their distressing coughing day and night continuously."[28]

Literary critic Susan Schweik notes that various municipalities passed laws prohibiting "unsightly" people from begging.[29] Applauding the recent passage of such an ordinance in its city, the *New Orleans Times-Democrat* explained that "street begging is confessedly a great nuisance, but when it is joined, as it has been in New Orleans, with the exposure of deformities, disease, and sores, it is simply unendurable." When the law proved ineffective, the paper complained: "The number of beggars is increasing in our streets and has already grown to be a nuisance. One old woman with a deep seated cancer on her face is a revolting sight."[30]

In short, incurability had new meaning at the turn of the twentieth century. No longer a synonym for chronicity, it now applied only to late stages of disease, when treatment had little prospect of success. The term also became associated overwhelmingly with members of socially disadvantaged groups, assumed to lack the moral character and physical constitution necessary to avert mortality. Nineteenth-century doctors and hospitals had often spurned patients labeled hopeless. Now,

Female ward for incurables in Blackwell's Island facility, 1916–1918.
Courtesy of the New York City Municipal Archives, NY.

changes in both medicine and the broader culture provided new justifications for denying their needs.

Tuberculosis Sanatoriums

The belief that people with tuberculosis were a menace to the uninfected population led to the establishment of special facilities for their care. In addition, growing numbers of TB doctors followed Peter Dettweiler, a German doctor, whose ideas were introduced to America by Paul Kretzchmar. "The smallest details of the patient's life," Kretzchmar wrote, should be "controlled by the supervising physicians and nothing of any importance . . . left to his or her judgment."[31] TB sanatoriums both segregated patients from the rest of society and subjected them to strict regulation.

The first U.S. sanatorium opened in the 1890s. By 1926, the nation had more than 500 facilities, with a total of 73,715 beds.[32] Both public and private institutions sought to serve a very different population from the inmates of municipal hospitals. Herman M. Biggs, the chief medical officer of the New York City Department of Health, drew a distinction between deserving and undeserving TB patients when he noted that he tried to reserve places at Otisville, his flagship sanatorium, for people with "incipient" cases who could be "restor[ed] to permanent usefulness in the community."[33] As his biographer explained, Biggs believed that "there were individuals . . . whose lives were so worthless to the community that it would be an unpardonable waste of public funds to give them the benefit of sanatorium care." Those "hopeless third stage cases, chronic alcoholics, and the persistently incorrigible" belonged in municipal hospitals on Blackwell's Island.[34] A few decades later, the director of the Washington State Sanatorium similarly conflated medical and social considerations when he asked, "Should incurables and patients with a prospect of recovery be treated in close association? Should derelicts, bums, streetwalkers, and other objectionables be treated in close association with our respectable, useful citizens?"[35]

Others justified the exclusion of patients with advanced disease in economic rather than moral terms. The resident physician at Barlow Sanatorium, a private Los Angeles facility, acknowledged that "to isolate any considerable number of advanced cases of an infectious disease is a most valuable means of helping to eradicate the infection by protecting those who are yet unaffected." Nevertheless, he concluded that "it is even more important, from an economic standpoint, to recognize and treat the early and curable instances of such disease, in order to preserve the life of the subject and prevent the long term of invalidism that is still the rule."[36] Another common argument was that the sickest patients inflicted higher costs because they required more care and could not work.[37] And some noted that patients approaching death disturbed those with more favorable prognoses. The medical superintendent of the Montefiore Home Country Sanitarium in New York argued that "advanced and far advanced cases" were "a source of depression and disappointment to incipient and hopeful cases. The sad sight of their more unfortunate fellow-sufferers, the relatively frequent occurrence of deaths, has been for the hopeful cases an unfortunate and premature warning of a *memento mori*."[38]

Discharge policies further helped to reduce death tolls. After stating that there had "been a careful elimination of those which were not deriving benefit from their stay here, and for whom the limits of the institution's usefulness has been reached" at the Montefiore Country Sanitarium, the medical superintendent remarked, "This action has been of the most humanitarian character, for by it the sanitarium has been able to accommodate a greater number of individuals whose conditions still warranted a confidence in the likelihood of their improvement or recovery."[39] New Jersey officials noted that both state and county sanatoriums sent home "those whose days are numbered and who beg to spend their last hours in familiar surroundings."[40] Because some "incurables" at Olive View, the Los Angeles County facility, could not return to their families, officials contracted with private rest homes to provide care.[41] As officials often acknowledged, the quality of care in these homes was seriously deficient. After visiting one rest home in January 1933, the director of the California State Bureau of Tuberculosis reported the death of "a woman from exposure because the pipes had frozen in the barn where she was being housed, so she was obliged to leave the building and walk over into another building to a toilet. She had a hemorrhage."[42]

Despite such measures, sanatoriums could never completely eliminate dying patients. New Jersey's goal was to "treat all cases in the minimal stage when recovery is particularly certain," but only 18.7 percent of those admitted to its public institutions were in the minimal stage.[43] A doctor at Barlow Sanatorium in Los Angeles noted in 1905, "Circumstances over which the institution has had no control, have made it necessary, in a number of instances, to admit patients whose condition gave not even the faintest hope of improvement." The sanatorium soon had a waiting list of nearly a hundred and gradually expanded its patient population to include many with terminal disease.[44] The staff physician of Olive View later wrote: "The pressure from the large number of patients needing treatment in Los Angeles County became so great that it was necessary to abandon the original plan of taking only early cases."[45] Of the sixty-six new patients admitted in April 1928, fifty had advanced TB. A nurse recalled that "it was really a terminal place." Many patients were "hemorrhaging and very, very ill" and "were coming because they knew it was the end for them."[46] And historian Barbara Bates noted that "more than half of the patients" at Pennsylvania's state sanatorium "had far advanced disease, living in cottages that were built for early cases."[47] Not surprisingly, the mortality rate of institutions throughout the country in 1926 hovered around 20 percent.[48]

Nevertheless, institutions prided themselves on their ability to cure, not on the care they delivered to the dying. Some held widely publicized homecoming days featuring the few former patients who held jobs. Many sanatoriums boasted of highly inflated "cure" rates.[49] Reports also emphasized the addition of new operating rooms and rehabilitation programs—but made no mention of the morgues and cemeteries the institutions were forced to establish.[50]

Sanatoriums attempted to conceal fatalities from patients as well as the public. As in hospitals, nurses removed dying patients from wards and failed to inform the other patients when death occurred. "Keep knowledge of death from other patients," read the rule book of a Virginia facility.[51] Staff members exhorted residents to remain optimistic and express only positive thoughts.[52]

But secrecy was difficult to sustain. When a state sanatorium in the Midwest tried to remove a boy's body under cover of darkness, several patients decided to "stay awake . . . to hear the wagon come to take him away."[53] The day after a woman died in the Washington State Sanatorium, the staff appeared "cheerful and brisk." Resident Betty MacDonald, however, "detected an ominous undercurrent" when she walked down the hall to the bathroom. There was a "furtiveness to the whispering," as patients spread the news the nurses had tried to withhold.[54] The sights and sounds of death figured prominently in memoirs of sanatorium life. Marshall McClintock, a patient in New York's Adirondack Mountains, recalled another patient who "looked like death. You knew that death was in her already. . . . You could see it in her face, in her eyes." She talked about returning home to her husband and children, but it was clear that she "would never go back to cooking and caring for those kids of hers, no matter what she thought."[55] Others described hearing patients choke to death.[56]

Black humor may have helped manage fears during the day, but dark forebodings shattered the bravado at night.[57] Assigned a bed on a narrow porch in a Wisconsin facility, Will Ross "felt close to the earth, closer than I wanted to be. It made me afraid, it made me think of death. It made me think of that vague eternity up back of the angry clouds that were pouring out their heart on the earth."[58] MacDonald thought she saw the "evil face" of death "shuffling up and down the corridors in the night. . . . Up and down the halls he went, never hurrying, knowing that we'd wait for him."[59]

If patients witnessed events they might have preferred not to see, family members were often absent. Will Ross received few visits because his "family lived an inconvenient train trip away."[60] Poor relatives were especially unlikely to travel. Urging that state and municipal facilities be located close to population centers, a public health nurse wrote: "No family will willingly send one of its members to die among strangers, too far from home to be visited. . . . It is not always understood what railroad fares mean, or how little accustomed some people are to even short journeys."[61] But the fresh air believed to be vital to healing was available only in rural areas far from inner cities. Otisville was thus located in the Catskill Mountains, seventy-five miles from New York City. Travelers had to pay for at least one night's accommodation in addition to the train fare.[62] Public transportation did not reach Olive View, in the northern part of Los Angeles' San Fernando Valley; instructions for traveling from downtown stated that people should take a bus or street car part of the way and then proceed by "private conveyance." Without the freeways that would later span Los Angeles, even car owners had to spend half a day traveling to the facility from some parts of the county.[63]

Children could have no contact with parents dying in sanatoriums. In an oral history interview Emilia Casteñeda de Valenciana recalled that although she and her brother went with their father to visit their mother in Olive View, "we couldn't see her. We had to keep our distance from the building. We used to spend our time just playing around on the property, but not too close because they didn't allow us to even go close to the porch where you could at least put your head against the screen door to peek in at your relative. We weren't allowed to do that because they were afraid that we'd catch the disease."[64] Emilia also noted how she learned of her mother's death:

> My mother died on a Sunday morning. I was making my First Communion. That's why I remembered the date. Since we didn't have a phone, we had given a neighbor's phone number to the sanitarium. Who could afford to have a phone in those days? . . . Anyway, the sanitarium had tried to call the neighbor, but she was gone somewhere for the day, and she didn't get home until late in the afternoon. My mother had died in the morning. . . . I didn't find out until after mass that my mother had died. I remember one of the girls in the neighborhood first told me that my mother had died, which was a shock, especially on such a happy occasion. When something like this happens happiness turns to sadness.[65]

Cancer Hospitals

Like tuberculosis sanatoriums, private cancer hospitals tried to restrict their clientele to "curable" patients. One of the earliest facilities was the New York Skin and Cancer Hospital, which opened in a small dwelling on Thirty-fourth Street in 1883 and soon moved to a more substantial building on Nineteenth Street. The hospital's first annual report noted that "many have been disappointed, and perhaps harm has been done to the institution by the enforced rejection of numbers of cases which have applied for admission." Those patients "were all in advanced stages, quite unfit for any operation, and beyond any hope of relief." Were they to be placed in the small wards of the hospital, they would "pollute the atmosphere to such a degree that the other beds could not be occupied." Efficiency also demanded their exclusion: "To fill up the comparatively few beds with hopeless, chronic, distressing, and offensive cases would, in the end, be to wrong a very much larger number of individuals who could with some certainty receive benefit in a short time."[66]

Perhaps recognition of the harm from that policy impelled the administrators to note the importance of caring for the sickest patients. The second annual report stated that the hospital was "organized to receive cancer cases of all forms and in every stage" and that one of its aims was "to minister to and soothe even hopeless cases."[67] A few years later, the facility established a country branch, primarily for patients in the late stages of disease. The 1907 annual report expressed pride in the care rendered to one dying woman. When "'Death the Friend' came gently to her," the hospital was left with a "memory that will not soon pass—of a little old lady, who endured pain without complaint, who had absolute trust in her Heavenly Father, and whose heart was filled to overflowing with gratitude for every kindness shown to her." But the country facility closed after a few years, and the 1907 report acknowledged that the case of this woman was exceptional. "Compassion" had "suggested that possibly the case might not be hopeless and on that ground she was received in this Hospital."[68] Although the hospital continued to care for some people with incurable conditions, most reports stressed the many operations performed, the large numbers of scientific studies conducted, and the high proportion of patients restored to "health and usefulness."[69]

New York Cancer Hospital (later General Memorial Hospital and then Memorial Hospital for Cancer and Allied Diseases) followed a similar trajectory. It was founded in response to the refusal of Woman's Hospital to accept cancer patients. As noted in chapter 2, when one physician admitted two cases in 1873, the Board of Lady Supervisors complained that the ward became "so very offensive" that two patients left and others protested "bitterly." Incorporated in 1884 with help of funds from John Jacob Astor, the cancer hospital opened in 1887 at the corner of Eighth Avenue and 106th Street. During the first year, the facility admitted 276 patients, 148 of whom received free treatment; the remainder paid full or partial charges. Although "no cases, however far advanced in the disease, were refused admission," successful applicants had to demonstrate "a reasonable prospect of relief by treatment."[70]

Paying patients were most likely to enter with late stages of disease. In 1891 the hospital reported that it had found it necessary "to restrict the number of free chronic cases to a single ward, not only because of the number of operations and the necessity for isolating septic cases, but to emphasize that the hospital is not an institution for incurables." According to the 1897 report, patients in that ward were "allowed to remain for six months, and in many instances for a year." The following year, however, the hospital announced the closing of the ward for free chronic cases "on account of the heavy expense involved."[71] Hospital regulations subsequently noted that patients with chronic disease were excluded from the wards—but there was no similar stipulation for patients occupying private rooms.[72]

Like administrators of the New York Skin and Cancer Hospital, those at Memorial acknowledged public pressure to care for patients with little hope of recovery. To do so, however, would "turn hospitals into almshouses and deprive them of the possibility of receiving" patients who could benefit from treatment. "While fully recognizing the claims of the advanced cases of malignant disease," the hospital believed its mission "would be defeated by devoting a very large part of its energies to this field."[73]

In 1920, the hospital announced that the average length of stay was now thirteen days (down from twenty-two in 1913) and that a social service department had been opened to place patients who no longer could benefit from treatment in other hospitals and institutions. A report two years later stressed that "the scientific interests of the hospital" demanded that the number of patients "in the terminal period of the disease" was now at a minimum.[74]

The patients excluded from Memorial had few alternatives. As one observer noted, many people with advanced cancer had a "constant necessity for the relief of pain, or for controlling hemorrhage or for employing special feeding methods, or for dressing wounds, or for other measures of relief." Because most acute hospitals did not accept such patients, who might live for months or even years before dying, they were forced to "seek relief in special institutions." Even then, the patient might wait weeks or months for a vacancy.[75] Although Memorial reported that its social service department gave "comfort and satisfaction" to everyone requiring follow-up care, a study conducted in 1928 noted that the facility had few places to which to refer the terminally ill patients it discharged.[76]

One option for those who qualified as indigent was the New York City Cancer Institute, established by the Department of Hospitals on Blackwell's Island in 1924. Official visitors a few years later reported that the facility lacked "decent and adequate facilities for the care of the sick."[77] The two-story brick building was "old and poorly equipped." The second floor had no porches to which patients could be moved. They also rarely went to the yard because there was no elevator; patients in wheelchairs had to be "carried up and down the stairs." The toilets were inadequate and dirty. Bathtubs were used to sterilize bed pans. "There were no utility rooms, no examining rooms, [and] inadequate diet kitchens." Although the report concluded that "good medical service . . . has been maintained under these incredible conditions," it also noted that the nursing staff was inadequate. Patients as well as investigators objected to the miserable conditions and "frequently left within a short time."[78]

From "The Shelter of the Doomed" to "The Treatment of Diseases"

Unlike the many tuberculosis sanatoriums and cancer hospitals that tried to exclude patients with little chance of recovery, homes for incurables actively sought their admission. The foundational stories of many homes begin in a similar way. A wealthy individual would become so disturbed by the lack of care for people with hopeless diagnoses that he, or more often she, would decide to establish a special place for them. Many homes opened with very few beds: there were two in the Washington Home for the Incurables in Washington, D.C., for example; five in the Virginia Home for Incurables in Richmond; and eight in the House of Calvary in New York City. As philanthropic and religious groups responded to appeals, the homes rapidly expanded.[79]

But the desire to ease the suffering of the dying did not always withstand the growing imperative to overcome death. Two New York facilities illustrate the divergent paths homes for incurables followed. Although both were founded under religious auspices and offered free care to indigent patients, they responded very differently to medicine's increasing focus on research and cures. Distancing itself from the prevailing medical concerns, St. Rose's Home for Incurable Patients clung tightly to its original mission. Montefiore Home for Chronic Invalids, by contrast, aligned itself with scientific advancement and gradually abandoned its commitment to patients with little prospect of recovery.

The founder of St. Rose's Home for Incurable Patients was Nathaniel Hawthorne's youngest daughter, Rose. Born in Lenox, Massachusetts, in 1851, she was educated in London, Paris, Rome, and Florence. In 1873, she married George Parsons Lathrop, subsequently assistant editor of the *Atlantic Monthly*, and began her own literary career, publishing short stories and a book of poems. Their son, Francis Hawthorne Lathrop, was born in November 1876 and died of diphtheria in February 1881. In 1891, both George and Rose converted to Catholicism, a highly unusual step for a woman who traced her ancestry to the Puritans and was the scion of one of the most prominent transcendentalists. Five years later, Rose decided to leave George and devote the rest of her life to caring for poor people with cancer.[80]

That dramatic decision often is explained in religious terms and as a response to the death from cancer of her friend Emma Lazarus.[81] Rose Lathrop later wrote: "I would not for the world intentionally give the impression that Emma Lazarus's death from cancer (since she was well off, & thus did not know of the suffering of cancerous poor) made me feel that I must found this work, though . . . it gave my mind some insight into the sorrow of cancer, & I was more able to entertain the idea."[82] The author Flannery O'Connor argued that there was a "direct line" between an incident in Nathaniel Hawthorne's life and his daughter's work. Visiting a workhouse in Liverpool, England, he had embraced a "wretched, pale, half-torpid" child who had followed him. Rose later commented that his story describing that incident represented his best writing.[83]

Less personal factors also must have influenced her. By the turn of the twentieth century, living and working among the poor had become an increasingly popular middle-class response to the varied social problems wrought by industrialization and urbanization. Lathrop would have been well aware of the growing number of settlement house workers and public health nurses in New York. The various activities of Roman Catholic sisters in that city undoubtedly served as an additional model. Some played a central role in establishing the city's Catholic hospitals; others founded orphanages, foundling homes, and old age homes.[84] And Lathrop could look to the example of the Home for Incurables already established in the Bronx as well as the fledgling facilities in other major urban centers.

After taking a three-month training course at New York Cancer Hospital in the summer of 1896, Rose rented rooms on Scammel Street in the Lower East Side. There, she spent her days caring for sick cancer patients in her own apartment and visiting them at home. Her literary background served her well. Using the name Hawthorne as well as Lathrop, she published numerous letters appealing for contributions in the *New York Times*. As donations poured in, she was able to move to larger quarters on Water Street early in 1897. At the end of the year, Alice Huber, a Catholic artist, joined her. After George Lathrop died in April 1898, the two women began the process of becoming Catholic sisters. In May 1899, they moved to Cherry

Street and opened St. Rose's Free Home for Incurable Cancer Patients (named after St. Rose of Lima), housing twenty patients. The following year, Lathrop and Huber formally entered the Dominican Order, Lathrop becoming Sister Mary Alphonsa and Huber, Sister Mary Rose. (For clarity, I will continue to refer to both women by their former names).

With help from wealthy friends, Lathrop purchased nine acres of land in Sherman Park (soon to be renamed Hawthorne), and in June 1901 she founded Rosary Hill Home, which could accommodate more than thirty patients. Although she occasionally complained about the constant need to raise money, she continued to attract funders. "My begging was very successful," she wrote to her nephew in 1919. "I find advertisements are sure to be printed, and the percentage on them is splendid—in this last instance, marvelous. In three papers the appeals cost about $250, & I received donations amounting to about $2,000!"[85] Lathrop lived at Rosary Hill Home until her death in 1926, while Huber assumed charge of St. Rose's Home; in 1912, that facility moved to a building on Jackson Street with room for a hundred beds.[86]

Lathrop's abundant published writings suggest that, like the founders of the alternative organizations that flourished in the 1960s and 1970s, she sought both to criticize mainstream institutions and to offer a model of exemplary human service.[87] At a time when most health care providers celebrated their growing scientific prowess, she insisted that her facilities were more homes than hospitals, focusing on care for sufferers rather than cure for disease. Shortly before opening St. Rose's, she explained that because there would be no surgery, teaching, or experimentation, she could dispense with that most "impressive feature of a hospital, its stream of physicians." One doctor would be sufficient to predict death and keep the patients as comfortable as possible. Unlike "his majesty, the young physician," she and her helpers would strive for humility rather than privileged status, living among the poorest of the poor and sharing their sorrows.[88] The lack of salaries further underlined the similarity to care delivered in the home.[89] And staff were forbidden "to show abhorrence or disgust at the sight of the repulsive ugliness brought about by cancer," to wear rubber gloves when tending patients, or to keep their dishes and utensils separate.[90]

In one important respect, however, Lathrop reinforced rather than challenged prevailing assumptions about terminal cancer. If she would not allow her staff to express disgust to patients, she never seems to have gotten over it herself. Indeed, her writings as well as her rules emphasized the "repulsive ugliness" of her patients. In an 1897 appeal for aid Lathrop remarked on welcoming a patient into her "poor quarters" on Water Street "in spite of the offensive cancerous condition from which she suffered."[91] A 1902 appeal noted that some patients "had half of their face gone, and enormous growths upon the side still remaining, with an eye gone, an ear half destroyed, a nose eaten away . . . (all being consumed still further day by day)."[92] Although Lathrop said little about how she assuaged the physical pain and existential agony of people close to death, the task of dressing wounds figured prominently in all her accounts. The work was thus "unpleasant," even "revolting."[93] She later described the first patient she encountered during her hospital training, a Mrs. Watson: "Her face was so terrible when wholly exposed that I trembled as with ague when first put to do the 'dressing' of it. This process was quite elaborate, so that I had my initiation with thoroughness."[94] After Mrs. Watson went to live with Lathrop at Scammel Street, Lathrop "never ceased to 'dress' her face."[95]

Alice Huber similarly emphasized the disgust advanced cancer aroused. Her memoir recounted her first visit to Lathrop's rooms on Water Street: "Everything about the place and the neighborhood, the untidy woman in the kitchen, the patients upstairs (who were certainly the most hideous we have ever had), seemed perfectly repulsive to me. I certainly did not want to stay. . . . I knew nothing whatever about taking care of sick people and felt great disgust to be near ordinary persons with sores, much less cancer." Rose Lathrop was "the only bright being in all that mass of ugliness and misery." Huber's visit the following week was no easier: "I loathed everything about the place, and still forced myself to dress a frightful sore that makes me sick to think of, even to this day."[96] Huber acknowledged that although she eventually devoted her life to care of the cancerous poor, she never fully conquered her sense of revulsion.[97]

While administrators and staff of many other institutions employed the language of disgust to explain their exclusion of dying cancer patients, Lathrop and Huber used it to emphasize their self-abasement and self-effacement. Just as medieval saints nursed lepers, the victims of the most loathsome disease of their time, so Lathrop and her staff lived in close proximity to dying cancer patients and dressed their gruesome wounds. Various admirers noted that Lathrop's unusually sensitive nature made her sacrifice all the more impressive, quoting her brother's comment that she had been "fastidious" as a child.[98] Others pointed to her pedigree to emphasize how much she had renounced. An 1899 letter to the *New York Times* soliciting funds for St. Rose's Home read: "We all know that if there is any nauseating, foul-smelling, despairing disease, cancer is almost the worst. . . . Yet Hawthorne's daughter has vowed her service for life to housing and nursing cancer patients. She herself attends to the patients every few hours to give whatever relief is possible, sleeping in a wall berth among them, so as to be ready for night duty."[99]

The rhetoric of disgust may have had a very different impact as well. As William Ian Miller writes, "Disgust . . . does more than register a simple aversion toward the objects of its focus. It degrades them in some way." The expression of disgust "always involves distance and superiority."[100] Emphasizing the offensiveness of patients with terminal cancer, Lathrop retained her place in the social hierarchy, undermining her humility even as she asserted it.

But if Lathrop maintained her discursive distance from the dying, she never deviated from her goal of serving them. The arrival of a gravely ill woman from the New York Skin and Cancer Hospital prompted this remark: "I wish somebody would stop those unstinted lies about cure which even that good hospital stoops to printing—for the sake of attracting subjects for experiment."[101] Like a growing number of critics of the use of sick people in medical research, she insisted that doctors too frequently sacrificed patient welfare to the demands of science.[102] Lacking a research component, her institution could focus exclusively on patient needs. Private letters to Huber occasionally described the special attention accorded patients as the end approached. When one woman seemed "to be dying," Lathrop "set watches by her night and day." A few months later, she noted that she had "been insisting on very constant attention" on another woman near death, "so that she does not suffer from various causes for a hour or two before any one appears."[103]

Spiritual guidance often was an important component of compassionate medical care. In 1915, Huber described a former boat captain who had been homeless for many years. Although originally a Catholic, he had drifted away from his faith. At St. Rose's Home, he "came back to his Church and the practice of his religion." As a result, "a peaceful calm filled his soul" and he "had a resigned, edifying death."[104]

After Lathrop's death in 1926, the Dominican Sisters of Hawthorne established four new facilities, in Philadelphia; Fall River, Massachusetts; Atlanta; and St. Paul. The sisters opened an additional home in Parma, Ohio, in 1956. By 1963, the homes served a total of 700 patients.[105] Visiting St. Rose's that year, British physician Cicely Saunders, the founder of the hospice movement, described the facility as "friendly and homelike with patient pictures and belongings in the right sort of untidiness in the wards."[106]

One reason Saunders may have singled out St. Rose's for special praise is that many other facilities had long abandoned their original missions. By the early 1930s, health professionals increasingly condemned homes for incurables for ignoring the therapeutic advances that could help to stave off mortality. A 1931 *Modern Hospital* editorial read:

> The term, "home for incurables," . . . is a reflection on the science of medicine, for who shall say when a patient is beyond all possible means of medical help? Some of the incurable patients of yesterday are the curable patients of today and it is therefore reasonable to expect that some of the incurable patients of today will be the curable patients of tomorrow. This cannot, however, be encouraged if we continue to segregate, isolate and forget them in "homes for incurables." The term itself is presumptuous because it creates an atmosphere of finality about the inmate and gives the impression that medical and social agents have passed a life sentence and in some cases a sentence of death, that may not be appealed.

The editors suggested that facilities reconstitute themselves as hospitals for chronic diseases.[107]

The Montefiore Home for Chronic Invalids provided a model others could follow. The Jewish philanthropists who founded that facility in 1884 chose the name as a "less terrifying" alternative to the original proposal, "Home for Incurables."[108] Nevertheless, the initial goal remained the same: "To afford permanent shelter in sickness and to relieve invalids resident of the City of New York belonging to the Hebrew faith, who by reason of the incurable character of the disease from which they may be suffering are unable to procure permanent medical treatment in any of the Hospitals or Homes."[109] Eighty percent of the first thirty patients were considered "absolutely incurable under all circumstances." Most had either paralysis or late-stage TB. Very few improved enough to be discharged; an even smaller number completely recovered.[110] "Ours is essentially a 'Home' to come to," wrote the Executive Committee in 1886. "When the inmates go, it is generally, alas! to that final resting-place which eventually claims all as its own." Discipline was thus far more lax than in the typical hospital, which controlled its patients "with almost military regularity." In order to "make smooth" the final "journey," "many whims and caprices have to be respected."[111] The following year the committee wrote that the staff "become so well acquainted" with the patients, "that we are soon familiar with their peculiarities and little weaknesses of temperament; and knowing just how to humor them, a bond of sympathy is established which tends in no small degree to make a happy household."[112] In 1893, the average length of stay was more than a year.[113]

One historian argues that lay control "remained stronger for a greater length of time" at Montefiore than at many hospitals. Because cures were rare, "the doctors did not seem to possess great power."[114] Yet, the first chief of staff, Simon Baruch, relentlessly tried to align the facility more closely with acute health care institutions, "constantly," in the words of one observer, emphasizing "the treatment of diseases as opposed to the shelter of the doomed."[115] A Polish-Jewish immigrant, Baruch graduated from the Medical College of Virginia in 1862 and then served as a surgeon in the Confederate Army. Returning to the South after the war, he became president of the South Carolina State Board of Health. He moved to New York in 1881, a few years before Montefiore opened.[116] Deriding staff members who derived emotional satisfaction from the "tender care of the incurable sick," Baruch used the construction of a larger building in 1886 as an occasion to express his hope that the institution's benefits would "accrue not only to those whose ebbing lives are our care, but also to the interests of medical science."[117] Baruch's abiding faith in the curative powers of hydropathy distanced him from some of members of the medical community. Nevertheless, he continually pressed the directors for new operating rooms and research laboratories throughout his twenty-year tenure.[118]

At a ceremony to mark the groundbreaking for a new building for the hospital on Gun Hill Road in the Bronx three years later, one speaker looked backward, declaring that Montefiore's primary goal was "to lessen the burden that many unfortunates have to bear—incurable disease. We establish a home for them and try to make what remains to them of life as pleasant and free from suffering as lies within human power."[119] But the balance of power already had begun to shift. The address of the current chief of staff was more in tune with Montefiore's new orientation, noting the need for skillful as well as humane treatment and stressing the institution's commitment to scientific development and research.[120]

In 1914, the board of directors changed the name to the Montefiore Home and Hospital for Chronic Disease. Shedding the word "Home" in 1921, the board explained that the "progress of medicine in the future lies in the study of chronic diseases" and that the institution had "a unique opportunity to be a leader in this advance."[121] More emphatically, the authors of the facility's fiftieth anniversary volume wrote that the change signaled "the decision never to accept a diagnosis of incurability as final."[122] The appointment of cardiologist Ernst P. Boas as chief of staff in 1921 further confirmed Montefiore's new direction. The son of the celebrated anthropologist Franz Boas, Ernst graduated from Columbia University's College of Physicians and Surgeons in 1914 and then interned at Mt. Sinai Hospital. After serving in World War I, he practiced privately while teaching physiology at his alma mater.[123] During his seven years at Montefiore, he organized both the medical staff and patients into distinct groups and completed its transformation into a major scientific institution.

Montefiore's admissions criteria changed along with its mission. In 1902, when Montefiore could admit only 521 patients (40 percent of the 1,330 people who applied), the Executive Committee noted that justice demanded that the hospital "should first consider the applicant who needs constant medical treatment and nursing" and "particularly such cases who have a chance for permanent improvement."[124] And many patients no longer remained until death. In 1914, the directors wrote that "with the advance of medical science," it had "become increasingly possible to discharge a very considerable percentage of the inmates after shorter or longer stays, often entirely cured—and more frequently vastly improved—and to make room for others entitled to the benefits of the institution."[125] By the late 1920s, the facility admitted people for three-month periods and discharged those who failed to show sufficient improvement.[126] Although the stay was still exceedingly long by the standards of acute-care hospitals, the new policy represented a radical change for an institution that once had sought to provide "permanent shelter."

Montefiore did not completely abandon the population it previously had served. In August 1921, it opened the Schiff Pavilion, where homeless patients requiring only custodial care could remain "for the rest of their lives."[127] Nevertheless, the institution now devoted only a tiny fraction of its resources to that clientele. As the director stressed in 1931, the former "home for incurables" had "evolved into a hospital for chronic diseases."[128] After World War II, Montefiore gradually became a general hospital, treating acute illnesses as well as chronic ones.

Public Chronic Care Hospitals

We saw in the previous chapter that private hospitals frequently sent people with chronic illnesses to public institutions. But dumping did not end when patients reached the public system. One historian writes that after receiving chronically ill patients from private facilities, public hospitals "shipped them as quickly as possible to isolated hospitals that cities had established for long-term care in out-of-the-way places—on nearby islands, as in Boston and New York, or in the far suburbs, as in Philadelphia, Chicago, and Washington."[129] Like municipal hospitals, public chronic care hospitals typically began as appendages to the almshouse, but they were much slower to acquire independent identities.[130] Well into the twentieth century, the chronic care hospitals in both Boston and New York remained adjacent to city almshouses and under the same administrative control.[131]

Although some patients in chronic care hospitals undoubtedly had stable conditions that were expected to persist for weeks, months, or even years, a high proportion were gravely ill. In 1903, the visiting medical staff of Long Island Hospital, Boston's chronic care institution, complained that the method used to transport patients from the city inflicted "discomforts and hardship, if not actual suffering" on patients who were "often in the last stages of disease." Delays in reaching the island were not only "inevitable" but also dangerous for the very sick. Such "inadequate provision for their comfort would not be tolerated in hospital wards" and "should not be during the trip."[132]

By 1900, the chronic care facilities in New York included not only City (formerly Charity) Hospital but also Metropolitan, first established on Ward's Island in 1889 and transferred to Blackwell's Island thirteen years later. Soon after that move, Metropolitan established a tuberculosis infirmary, which rapidly grew larger than the parent institution.[133] Explaining why 624 of the 2,415 patients died during its first year of operation, the infirmary administrators wrote, "It must be remembered that . . . many cases come here in extremis and die within a few days of entrance."[134] Largely as a result of the reforms instituted in the late 1890s and early 1900s, transportation to Blackwell's Island greatly improved. Nurses rather than workhouse inmates were responsible for patients throughout the boat trip, and patients on stretchers traveled in a special, heated room.[135] Nevertheless, both City and Metropolitan Hospitals continued to direct the same criticism toward Bellevue that it

Patient arriving by ambulance to a Blackwell's Island facility, late 1890s. Courtesy of The Museum of the City of New York/Art Resource, NY.

leveled against private hospitals—too many dying patients were transferred, and the rigors of travel often accelerated the end.[136]

Doctors and nurses at Blackwell's Island facilities appear to have found few rewards in caring for the patients their colleagues shunned. A 1928 study reported that "pressure from the medical and nursing staffs" at City and Metropolitan Hospitals "led to a change of policy in 1910," permitting the entry of at least some acute-care patients.[137] A physician recalled that after the increase in admission of acute cases at Metropolitan Hospital, "nurses and doctors alike had their enthusiasm aroused."[138] When city officials proposed transferring the acute-care patients back to other facilities to relieve overcrowding on the island, the commissioner of public charities protested that it was "necessary to maintain an acute service in a chronic hospital in order to secure and retain attending physicians of ability and merit." Before the entry of acute-care cases, "the medical service in these hospitals was far from satisfactory." The great improvement stemmed "almost entirely from the introduction of acute cases." Were the city to "discontinue the acute service at the City and Metropolitan Hospitals," they would "lose the services of [the] best attending physicians, which would seriously cripple the medical work."[139]

Report after report complained about the deplorable conditions in chronic disease hospitals. A 1903 investigation of Boston's Long Island Hospital found extreme overcrowding as well as numerous examples of neglect and abuse. In May of that year, a nurse administered an overdose of strychnine to four patients, two of whom died.[140] The reporter for a 1908 *New York Times* article wrote that "the top floor of the woman's building" of City Hospital, housing "patients in the last stages" of tuberculosis, had "so many cots in the room that anything like privacy for the patients was out of the question. There was not a corner where the dying women could be taken for their last hours."[141] In 1911 the New York City Charity Organization Society issued a report castigating the city for the "disgraceful" overcrowding in Metropolitan's TB infirmary. Beds "regularly lined" the halls, and many patients were forced to sleep on mattresses on the floor.[142] Four years later, the commissioner of public charities wrote to the mayor: "I desire to take advantage of this opportunity to go on record again squarely as stating that the condition of overcrowding existing . . . in certain wards of Metropolitan Hospital constitutes a disgrace to the City of New York."[143] The buildings were squalid. As the commissioner of public charities wrote in 1911, his department had "for years" "repaired and altered and put up with its old structures, many of which, because of great age and hard usage, are after all little more than makeshifts and apologies."[144]

Small budgets prevented administrators from hiring sufficient numbers of nurses. According to the *New York Times*, a former commissioner of public charities told

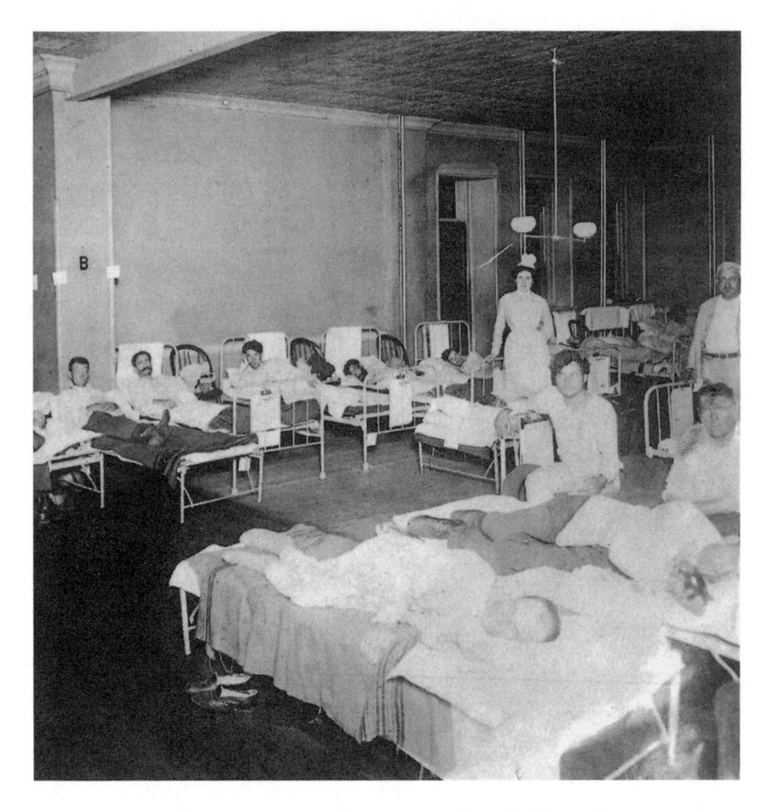

Male tuberculosis infirmary, Metropolitan Hospital, probably early twentieth century. Courtesy of the New York City Municipal Archives, NY.

a meeting of taxpayers in 1910 that "the conditions among the sick poor on Blackwell's Island were wretched" because poor funding forced the hospitals to employ "inefficient help." He used the following example to prove his point: "One of the physicians had instructed the nurse to give a woman patient a bichloride bath. In a short while the bath was prepared and the patient put in the water, carefully heated to the proper temperature. Then the nurses gave her a dose of bichloride of mercury, a corrosive sublimate, and the Coroner, of course, had to be called."[145] In 1915, the commissioner of public charities noted that City Hospital had approximately half the nurses "necessary to provide reasonably efficient nursing."[146]

Observers occasionally suggested that the nature of the clientele as well as inadequate staffing levels explained the poor quality of care in chronic care hospitals. A visiting physician at Long Island Hospital argued that "the conditions of nursing there are extraordinarily difficult, in the sense of being very discouraging to the young woman there—chronic patients who under normal conditions would be rather difficult old people to take care of, who are heavy to change, and who are not always absolutely cleanly in their habits." Another doctor at the facility agreed: "The work of nursing this sort of patient is the hardest work a nurse can have, the least interesting work that a woman is called upon to perform, and very great credit is due to the nurses at Long Island for the patience and devotion they have given to this altogether unpleasant work."[147]

The isolated settings of chronic facilities discouraged family visits. Urging that Boston's chronic disease hospital and almshouse be moved to the mainland, administrators wrote in 1923: "The present site . . . on Long Island shuts off the inmates almost entirely from contact with other people. Friends and neighbors who would be glad to visit them find it almost impossible because the boat trip there and back takes half a day, time which is difficult to find since most of them are working people. The average number of visitors per person is about one a year."[148] A woman who sought to prevent her husband's transfer from Bellevue to Metropolitan Hospital explained to a social worker that the trip would be too far for her and the children; she wanted him to remain where they could visit him almost every day.[149] And delays in informing relatives of transfers could prevent them from attending deaths. One woman did not learn that Bellevue had sent her husband to Blackwell's Island until a few days after he died.[150]

In 1935, S. S. Goldwater, then commissioner of the New York City Department of Charities, wrote: "Many communities inconsistently spend millions for the construction and maintenance of hospitals for the acutely ill while begrudging even small sums for the care of chronics. Neglected, dilapidated, hopeless in outlook, the chronic hospital often arouses a feeling of repugnance if not one of downright disgust."[151] Because the words "repugnance" and "disgust" commonly applied to people with advanced diseases, some readers may have assumed his remarks included the inmates as well as the facilities that housed them. His vivid language, however, helped to call attention to the miserable conditions patients continued to endure throughout the country.

Unintentional Refuges for Incurable Patients

Inadvertently, both almshouses and private old age homes represented major sites for the care of incurable and dying patients. Partly because reformers removed other groups from almshouses, sick and disabled people constituted a very high proportion of the inmates.[152] Many arrived in desperate straits, having exhausted all other resources. And poor food, unsanitary conditions, and the lack of adequate medical care accelerated death while intensifying suffering. A 1930 article in the *American*

Journal of Nursing noted that "the reports of the care of the sick in many almshouses give such a picture of neglect, indifference, and poor management through ignorance as to be almost medieval."[153] A study by the U.S. Bureau of Labor Statistics found that the matron, typically the wife of the superintendent, was "chiefly, often solely, responsible for the care of the inmates in illness, a responsibility which, considering the usual age of the patients, may mean life or death."[154] Investigators from the Woman's Department of the National Civic Federation visited one small almshouse where an elderly woman "was dying literally by inches from a complication of organic diseases." She was incontinent and had "to be cleaned and dressed and her bedding renewed day and night." The matron provided the only nursing as well as all the cooking and cleaning. As she told the investigators, the inmates "don't live long after they are brought here—this is their last stop."[155] The North Carolina State Board of Charities and Public Welfare found an elderly man dying of stomach cancer in another small almshouse "without attention."[156]

Historians Carole Haber and Brian Gratton demonstrated that the death toll mounted especially rapidly in segregated facilities in the South.[157] With a much smaller per capita budget than the white almshouse, the Ashley River Asylum in Charleston, South Carolina, had unusually dismal conditions. The city's 1924 annual report indicated that the black facility "has no water and sewage, and no electric or gas lights. The water which the inmates drink comes from an open well or an old cistern. The buildings are dilapidated, old frame structures, heated by wood stoves, lighted by lanterns and lamps."[158] Severe depression as well as lethal conditions may have weakened immune systems and led to death. Forty percent of the Ashley River inmates died annually, four times as many as at the white institution.[159] Dying patients enjoyed little comfort. The 1924 report noted that "the inmates were sleeping on straw mattresses, most of them on beds that had no springs."[160]

Like many other reformers, New York City commissioner of public charities Homer Folks tried to upgrade the Blackwell's Island almshouse, renaming it the City Home for Aged and Infirm in 1903.[161] Twelve years later, however, John A. Kingsbury, the current commissioner, wrote that serious overcrowding prevented the facility from providing "a reasonable standard of decency in the care of its dependent aged and infirm population."[162] A letter to the mayor from a New Yorker who had visited the almshouse the following year provided anecdotal evidence of that lack of decency. An elderly woman who had died from "a compilation of diseases" had never been "allowed to go to bed until she was actually dying. That was the rule of the place. The poor creature, dying by inches, was allowed, however, to put her head on the bed."[163] A 1933 study concluded that the home lacked "adequate medical facilities and even [fell] below the accepted standard of custodial care for the chronic sick."[164]

A major goal of private old age homes was to rescue people from the almshouse. The Hebrew Home for the Aged in New York sent "committees . . . to the city hospitals on the islands, and many of the aged were brought to the Home to spend their last few days amongst people of their own faith and race."[165] Homes established under Protestant auspices typically restricted their clientele to people considered too respectable to associate with almshouse inmates.[166] Many facilities also stressed their distinction from homes for incurables, demanding that applicants submit proof that they did not suffer from irremediable disease.[167] Directors could not, of course, ignore the frequent occurrence of mortality. "Although the homes were designed for the living," historian Carole Haber writes, "death dominated the pages of their annual reports. If few residents had died during the course of the year, the homes often credited their sanitary conditions, good food, kind attention, or medical care. If large numbers had passed away, however, the home expressed no great concern; this, after all, seemed a reasonable expectation."[168] But the directors expected the period of infirmity to be relatively brief. " 'Tis a wink of an eye, 'tis the draught of a breath," wrote the director of the Boston Home for Aged Men in 1869, "from the blossom of health to the paleness of death."[169]

As chronic conditions became the major cause of mortality, fewer and fewer residents died swiftly and suddenly. Lacking other options for care, residents could not be discharged when their bodies and minds began to decline. A 1928 New York City survey concluded that "without intending to do so, the private homes for the aged have assumed a large part of the community's responsibility for the care of the chronic sick." The primary causes were "general physical deterioration connected with senescence" and "circulatory diseases." Medical resources, however, were sparse. "The larger private institutions sometimes maintain[ed] a hospital department for guests who become ill; but a majority of the sixty private institutions . . . were without facilities for proper care of the chronically ill."[170]

Dying at Home

The dearth of institutional beds for terminally ill people typically escaped public attention, but a few charity workers, public health nurses, and physicians tried to arouse concern. One of their central arguments was that very sick people posed unbearable burdens on family members and thus needed to be in institutions. Tenement houses were described as the worst possible places for the delivery of care.[171] "Imagine the discomfort to the patient, as well as to other members of the household, when forced to live out the remainder of life within the narrow confines of a tenement-house home!" wrote William Seaman Bainbridge, a professor of surgery at New York Polyclinic Medical School and Hospital. "Imagine the cross which the care of such a patient, no matter how dearly beloved, inflicts upon the busy mother of the family, when her heart is already heavy with its weight of poverty and hardship, and her hands are overburdened with a multiplicity of household duties!"[172]

Because chronic disorders tended to accumulate with age, Ernst P. Boas emphasized the difficulties elderly people inflicted:

> The parents become such a burden to the younger generation that, after a while, the children welcome any means that will enable them to be rid of them. Not infrequently it is the son-in-law or the daughter-in-law who will not tolerate the presence of the invalid in the home. The patient himself may be so exacting in his demands and so self-centered, that home care becomes impossible. . . . Night after night, a mother, a father, a daughter or a son may have his sleep disturbed by the calls of the patient; the whole atmosphere of the home becomes subdued, the children lose their spontaneity, and life assumes a drab color.[173]

Others focused on the problems of caring for people with specific diseases. The New York Committee on Relief of the Committee on the Prevention of Tuberculosis explained why institutional placement was essential for all indigent TB patients, especially those close to death: "Applicants for charitable aid because of tuberculosis have the disease in such marked form, that the main consideration, from a medical point of view, is one of preventing them from infecting the other members of their households. The attempt at cure or arrest of the disease is not infrequently further complicated by the ignorance, bad habits and the poverty of the applicants." The committee concluded that "relief in the homes of consumptives in a far advanced stage" was "an unsatisfactory and dangerous substitute for the isolation provided by hospitals."[174] Bainbridge spoke to the hardships cancer patients imposed: "In pronounced cases, with visible ulcerations, and odors that cannot be disguised, the patient must be fed apart from the rest of the family; his clothing must be laundered separately; the ulcerating surfaces must be dressed and the dressings disposed of; while the suffering of the afflicted one must be witnessed alike by the old and the young of the household."[175] Despite repeated references to the burden that very ill patients placed on families, women represented the overwhelming majority of caregivers in the past, just as they do today.

Some comments about the difficulties imposed by the sick also may have represented condemnation of domestic healing. During the early twentieth century, health professionals relentlessly asserted that they alone had the knowledge and skills necessary to provide adequate care.[176] Nevertheless, some criticisms of home care had merit. The case files of the New York City Charity Organization Society document the difficulties of tending tuberculosis patients in tenement houses. To prevent the spread of germs, visiting nurses instructed families to isolate tuberculosis patients in separate rooms and maintain scrupulous cleanliness. But many apartments were extremely small. Only the rare family could afford additional beds. And the dilapidated condition of buildings, the soot and ashes produced by coal-burning stoves and kerosene and gas lamps, and the serious overcrowding all made dirt unavoidable. Cleanliness was especially difficult in the final stages of disease, when patients coughed and vomited frequently, soiling themselves, their beds, and sometimes even the walls around them.[177]

Other evidence reminds us that as news about the communicable nature of tuberculosis spread, growing numbers of caregivers throughout the country were well aware of the dangers of contamination. Martha Shaw, an Iowa woman, accompanied her tubercular husband, Johnny, a postman, when he traveled to Los Angeles to regain his health in 1893. There, his condition rapidly deteriorated, and he no longer could work. Forced to find employment as a waitress, Martha confided to her diary that although the job was hard, it was also "a blessing, in that I do not have to be so much with Johnny and run the risk of taking consumption, for he coughs dreadfully and the smell from his body is sickening: smells like his body was death." The entry ended with the wish that she would "not have to sleep in the same room with him."[178]

Twenty years later, Eve Matthews cared for her sister Abigail, dying of tuberculosis in Sheffield, Vermont.[179] Because Abigail's condition was far advanced, both the Vermont sanatorium and Dr. Edward L. Trudeau's private facility in Saranac, New York, refused her admission. The two young women thus spent the winter before Abigail's death first in a tent and then in a shack on the property of relatives. A nurse who visited reported that Eve "was busy every minute working for Abigail's comfort."[180] Eve noted she could find no relief from care simply because everyone "was afraid of the disease."[181]

A 1914 letter from Minnie Radamacher, in Long Beach, California, to an Illinois relative described the difficulties of caring for her sister Ida, dying of cancer. Minnie helped with the housework, but Ida's daughter Clara provided the bulk of the care. "We are all well. Except poor Sister she is still suffering with that dreded disease the canser and is getting worse all the time," Minnie wrote. "Poor thing, she is nothing but skin and bones and the terrible odor that comes from her wound is almost unbearable and still Clara has to clean it 3 and

4 times a day. . . . It takes all Clara's time to wait on her Mother by day and by night."[182]

Not everyone, however, agreed that family care was uniformly burdensome. Administrators of medical institutions occasionally justified the discharge of incurable patients by stressing the importance of allowing them to die at home. During the 1950s, many health professionals advocated expanding home care services for indigent patients; some emphasized the benefits that accrued to families as well as patients when death occurred outside institutions.[183] Such rapid changes in the prevailing discourse suggest that some arguments about the burdens of care may have been rationalizations for other ends.

And evidence indicates that even in very poor households, caring for very sick patients often was far more than a set of exhausting tasks and demands. The case files of the New York City Charity Organization Society are filled with conflicts between charity workers, who sought to send seriously ill and dying people to institutions, and relatives, again primarily women, who wanted to keep loved ones at home. Resistance to institutional placement was especially fierce when patients were close to death. Under pressure to hospitalize her husband with advanced tuberculosis, one woman said that she "never had been separated from him and could not bear to let him go now that he is so ill."[184] Women stressed not just their own sorrow but also that of the patients. When the Charity Organization Society urged a 14-year-old girl to enter a city hospital six months after leaving Otisville, the municipal sanatorium in the Catskills, in 1912, the mother argued that the girl's extreme loneliness at the sanatorium showed that another separation from home would hasten her death.[185]

Six years later, the society engaged in an extended struggle with an Italian immigrant couple over the care of their 12-year-old daughter, another TB victim. When the girl was hospitalized against her parents' wishes, an aunt protested that "the child will grieve herself to death because she is separated from her family." After the girl was transferred to Metropolitan Hospital, her mother argued that she would die from homesickness if compelled to stay. Shortly before the girl died, two months after her admission, her cousin wrote, "I think it is a sin to have a sick child suffer the way she is."[186] We also can recall the statement of a public health nurse in the 1930s that families did not willingly send members to distant sanatoriums "to die among strangers." Despite the serious, occasionally calamitous consequences of caregiving, many relatives wanted terminally ill patients to remain at home.

But families were not always available. Growing numbers of "unattached" individuals congregated in turn-of-the-century cities. And even in rural areas, where institutional beds were especially scarce, social networks often were either nonexistent, overstretched, or frayed. The Thomas Thompson Trust, a charitable agency providing assistance to poor women in and around Brattleboro, Vermont, and Rhinebeck, New York, received numerous requests from elderly invalids bereft of support while facing serious chronic illness between 1915 and 1940. Reports by the agents who conducted investigations describe the women's situations.

One was 88-year-old Becky Masterson, who had remained in the hospital for a long period after breaking her hip.[187] The hip still had not healed, but now her bed was needed for an acutely ill patient, and it was not clear where she could go. Her only relatives were a sister-in-law, who was also "elderly and in poor health," and a poverty-stricken brother whose wife had just died. Another was Constance Spalding, a "frail old lady" living alone in a tiny apartment. She could not "get herself her meals properly," was "very trembly," and was "likely to fall when there would be no one to see her." Her one surviving family member was the granddaughter she had raised, now teaching in another town, too far to come on weekdays. Constance died the following year of a broken hip. Mattie Jones was "stricken with a shock" while working at an old age home. Her left side was "entirely paralyzed and she [was] entirely helpless, being too weak even to feed herself at all." The doctor thought she could have another "shock" at any time. While Mattie was still in the hospital, the sister-in-law with whom she had lived died suddenly of ptomaine poisoning, so there was nowhere for Mattie to go even if she recovered. Ingrid Hoover had "hardening of the arteries which has upset the heart action so that she has frequent attacks of severe pain any one of which she is likely to pass away [from]." Although Ingrid was seventy-three, she was "as feeble as a person 80 or over." The sister with whom she lived was "now very much worn with the care." The daughters were "not able to give any money or to take her into their homes as they have all they can swing."[188]

Patients might lack adequate care even when sharing homes with female relatives. Many were alone all day while caregivers worked. Forty-year-old Lucy Harkness was considerably younger than most other applicants to the Trust. In the final stages of cancer, she was in constant, intense pain. The nights were a special "terror to her" because her mother was deaf. Although the doctors had warned Lucy that she could have a hemorrhage at any time, she knew that if anything happened to her during the night, her mother would not be able to hear her.[189]

Care or Cure

In chronic disease institutions, as in acute care ones, the demand to avert death gradually trumped more traditional responsibilities for patients at the end of life.

Although neither tuberculosis sanatoriums nor cancer hospitals could expel all dying patients, both gave preference to people with curable conditions. Public chronic disease hospitals had to accept all who applied, but in New York doctors and nurses campaigned for the admission of more patients with problems that could be quickly and successfully resolved. And many facilities originally established as homes for incurables gradually shed their commitment to terminally ill people. By 1906, the Montefiore Home and Hospital for Chronic Disease took the position that "alleviation of suffering and relief of distress are surely noble work, but the broader view into the future gives its possessor the desire to prevent all these evils and stimulates his best efforts to discover a way to accomplish this."[190]

The health reform movements of the 1920s and 1930s neither corrected medicine's tilt toward acute diseases nor raised the visibility of dying patients. Reporting in 1932, the independent but extremely influential Committee on the Cost of Medical Care concluded that because the expense of many health care services was prohibitive to the poorest Americans, resources were "inequitably distributed."[191] As the Depression deepened, many people who had considered themselves middle class could not pay for care. Because most Americans continued to associate chronic disorders with indigent inmates of state and municipal institutions, the overriding concern of advocates for compulsory health insurance was the cost of acute illnesses.[192] The failure of the drive for a compulsory insurance program led to the establishment of Blue Cross, the first prepaid hospital plan, which also focused exclusively on acute conditions.[193]

During the same decades, a small group of doctors, statisticians, and public health officials argued that the growing prominence of chronic disorders demanded a shift in medical priorities. Criticizing "the fetishism of the acute," they campaigned for increased funding for research into the causes of chronic afflictions and better care for the victims.[194] The dying were not part of the group's concern. Whereas turn-of-the-century physicians had sought to restrict the term "incurable" to the final phase of chronic disease, those advocates sought to eliminate the label entirely. "As medicine progresses, the conception of incurability is constantly changing," declared a 1933 report. "When a doctor calls a patient incurable, he is confessing his ignorance of the nature of the disease."[195]

A leading member of that group, Ernst P. Boas continued to call attention to what he called "the unseen plague of chronic disease" long after leaving Montefiore. Although he condemned the poor quality of care in various types of chronic care facilities, he directed some of his harshest attacks toward homes for incurables, the type of institutions from which he had tried to distinguish Montefiore: "Quite aside from their inhuman name which kills the last hope of any unfortunate who gains admission, their whole medical policy is based on the assumption that their patients are hopelessly and incurably ill, that any medical treatment beyond that which may be needed to keep them fairly comfortable is extravagant and unnecessary." "Furthermore," he continued, "most 'homes for incurables' do not take the pains to make a careful selection of cases."[196] If Rose Hawthorne Lathrop had condemned hospitals for experimenting on the dying and publicizing illusory cures, Boas argued that facilities like hers viewed patients "as human derelicts," unworthy of "constructive medical treatment."[197] The clash in perspectives was partly a controversy about whether to give priority to care or cure. Emphasizing the need to minister to the dying, Lathrop said little about the possibility that some of her patients might benefit from further treatment. Pointing to the continually expanding opportunities for recovery, Boas criticized the fatalism inherent in homes for incurables but ignored the need to accept what ultimately could not be changed.

"All Our Dread and Apprehension"

"Weeks of shocked apprehension, anxiety, and strain," wrote music composer Dorothy Smith Dushkin in her diary on December 6, 1959. Born in Chicago in 1903, Dorothy had graduated from Smith College in 1925 and five years later married David Dushkin, another musician, with whom she established a famous music school and summer camp. Now Amanda, the youngest of their four children and the one "with no problems in her well-adjusted & exuberant life," was "suddenly a victim of physical abnormality." A Smith College sophomore, Amanda had complained of "recurrent pains" in her side and back. Because X-rays at the college infirmary had revealed a probable tumor, Dorothy had taken the girl to Massachusetts General Hospital in Boston, where a "top-flight" chest surgeon, Dr. Gordon Scannell, performed surgery two days later. After what "seemed like a frighteningly long operation," Dorothy learned that the tumor, though not benign, was "probably of a low malignancy." The exact type would not be known for a few days, but it was definitely "not the dangerous type called Ewings." The doctor further reassured her that he "got every bit of it out & that was the most important thing."[1]

On March 27, 1960, Dorothy wrote again: the "most unlikely probability of tumor recurrence did happen." This time "the doctors left no doubt in our minds about the dangerous condition." Tests had shown that Amanda had "a rare case of explosive cancer" and that the disease had spread. Because surgery had been ineffective, the doctors would use two new treatment modalities—"super-voltage x-ray, but first five days of intra-venous dosage of new drug Actinomycin D." Although physicians had administered radiation to patients since the early twentieth century, the development of megavolt X-ray sources represented a major breakthrough. Actinomycin D was an extremely toxic antibiotic, which scientists recently reported could counter various malignancies when used in conjunction with radiation.[2] Amanda's physicians initially stated that they administered the drug "to make the x-ray more effective" but later acknowledged that it "was actually their only hope, as one area of the malignancy (behind the lung) could not be reached by x-ray." The outlook was poor. The specialist in charge of the case "bluntly" announced that Amanda had just a 1 percent chance of recovery and that she might die in a few months. The treatment side effects were horrendous—first, violent nausea and then when that abated, Amanda's "skin broke out, lining of nose and throat swelled and on top of it all she contracted the flu with a high fever for 2 days."[3]

Because Dorothy feared that new tests would confirm the dire prognosis, she stayed home while Amanda's father and sister took Amanda for a consultation. Soon David phoned to announce a "miracle." "The drug had been amazingly effective—had almost cleared up the diseased area. . . . The three doctors involved—surgeon, cancer specialist and x-ray therapist—were all astounded." Although no physician would give "absolute guarantee of cure," hope had been restored. [4]

"A month filled with . . . deep anxiety," Dorothy reported on January 6, 1961. At Thanksgiving Amanda's right eye had drooped and her vision worsened. Even before a biopsy at Mass. General confirmed the doctors' fears, the actinomycin D and X-ray treatments were started immediately. The drug "made her very nauseated again, but it had the same marvelously quick effect on the malignancy." Within a week, Amanda's eye was "back in place" and her vision normal.

Then on the night of February 26, Amanda was "much worse," Dorothy wrote the next day. Amanda had just reported pain and weakness in one leg. Although the results of the X-ray had not arrived, the doctors assumed that the cancer had returned and that Amanda would have to return to the hospital "to do battle again." Dorothy's faith in the medical regime had gradually dimmed: "How long her body can take the drastic injection of actinomycin D, I don't know," she wrote. "I don't believe the doctors do either." [5]

The news was much brighter on March 19: "Another reprieve for Amanda and a lifting of apprehension for me." An examination revealed that she could be "spared" the drug treatment. Because she needed only X-ray therapy, which could be administered in the Cooley Dickinson Hospital, located in Northampton, she could return to school.

Although Dorothy wrote little about her daughter during the following autumn, it is clear that she gradually realized that the end was near. On October 24, she noted the need to find a way to accept "the outcome the doctors expect." On December 12, she referred to Amanda's "desperate illness." And then, on January 5, 1962, she wrote from a room adjoining Amanda's in Cooley Dickinson. Now Amanda seemed to see or hear little. When the medication began to wear off, she "moves a little, moans &

forms words with her lips though the skin around her nose & eyes & mouth is grey and transparent, her cheeks have a delicate flush." A note early in the morning two days later recorded Amanda's death.

"On the Brink of Breakthroughs"

The Dushkin family's tragedy unfolded toward the end of a period of medical triumphalism. America's victory in World War II, abetted by the development of penicillin, radar, and the atomic bomb, had generated unprecedented optimism about the entire scientific enterprise. The war "taught one lesson of incalculable importance," the *Woman's Home Companion* reported in 1946. "The lesson: that with unlimited money to spend we can buy the answers to almost any scientific problem."[6] Both federal and private funding for medical research soared. The most spectacular result was the development of the Salk vaccine. One Salk biographer noted that after the announcement on the morning of April 12, 1955, that the vaccine had proved effective, "people observed moments of silence, rang bells, honked in brief periods of tribute, took the rest of the day off, closed their schools or convoked fervid assemblies therein, drank toasts, hugged children, attended church, smiled at strangers, forgave enemies."[7] Newscaster Frank Deford, a Baltimore fourth-grader at the time, recalled thinking, "We were safe again. . . . We had conquered polio."[8]

Despite the terror it aroused, polio was a relatively uncommon disease with a low case fatality rate.[9] Other researchers sought to vanquish the chronic diseases that contributed far more significantly to the nation's death toll. In 1946 Selman Waksman announced the discovery of streptomycin to cure tuberculosis. "Thus," he wrote twenty years later, "a disease that less than two decades ago was still regarded as the greatest threat to the health and life of man, a threat that hung over the heads of people like the sword of Damocles, has been reduced to the tenth position or even farther back, among the killers of human beings." Waksman took justifiable credit for saving many lives but ignored the decline in the tuberculosis mortality rate that had begun decades earlier as well as the advent of less toxic and more effective chemotherapeutic agents.[10]

By 1950, the leading cause of mortality was cardiovascular disease, which was responsible for 354 deaths for every 100,000 people. The second was cancer, accounting for 140 deaths per 100,000 population.[11] But research advances convinced scientists they were on the verge of controlling and possibly even eliminating both scourges. In 1944, Alfred Blalock, a Johns Hopkins surgeon, performed the first operation on a "blue baby" (a child whose congenital heart defect had restricted the oxygen supply and caused the skin to turn blue). A doctor who had assisted Blalock later noted that physicians, after receiving news of the successful operation, "altered their attitude and started to think of cardiac surgery as an essential component in the therapeutic armamentarium for disorders of the heart."[12] Other dramatic improvements in cardiac care came from the introduction of antibiotics, oral diuretics, and hypertensive medication and the development of cardiac catheterization, a surgical procedure enabling physicians to diagnose heart problems with far greater accuracy than before.[13]

The June 27, 1949, issue of *Time* magazine epitomized the widespread excitement surrounding the postwar cancer research industry. On one side of the cover was a drawing of a sword, symbolizing the American Cancer Society, slashing through a crab. The rest of the cover was devoted to a picture of a grave but confident-looking Dr. Cornelius P. Rhoads, the director of both New York's Memorial Hospital for Cancer and Allied Diseases and the Sloan-Kettering Institute. Memorial had changed dramatically since its foundation as the New York Cancer Hospital in 1884.With funds from the Rockefeller Institute, the hospital constructed a new twelve-story building on East Sixty-Eighth Street and First Avenue. Closely affiliated with Memorial, Sloan Kettering had been established in 1945 as the nation's first cancer research cancer research institute. *Time* dubbed it a "tower of hope."[14]

Rhoads used equally inflated language. A former army officer, Rhoads had served as chief of the army's Chemical Warfare United during World War II, when he had witnessed the ability of mustard gas to retard cancer cells. Now he argued that the scientists under his direction fought to destroy "an army of invaders of the body," using "atomic, chemical and biological means."[15] Although radiation therapy initially garnered far less attention in the United States than in Europe, its prominence rose in the wake of the explosion of the bombs in Japan and the new enthusiasm about the possibilities of atoms for peace. "Medically applied atomic science has already saved more lives than were lost in the explosions at Hiroshima and Nagasaki," claimed *Hygeia*, the American Medical Association's popular journal, in 1947.[16]

Chemotherapy, increasingly Sloan-Kettering's focus, aroused even more interest because it was a systemic, rather than a local, treatment and thus could have far more effect on metastasized disease.[17] Rhoads anticipated that drugs would be able to play the same role in cancer treatment that antibiotics played in the cure of infectious disease.[18] "Inevitably, as I see it," he declared in 1953, "we can look forward to something like penicillin for cancer, and I hope within the next decade."[19] The following year, Congress awarded the National Cancer Institute (NCI) $3 million to test various chemotherapeutic agents; and by 1958, scientists at various institutions annually screened approximately 20,000 compounds. Less than two years before Amanda Dushkin received her first actinomycin D treatment, the director of NCI's Cancer Chemotherapy National Service Center announced that antibiotics ultimately would provide a cure for some of the most common and intractable cancers.[20] "Cancer—On the Brink of Breakthroughs," proclaimed the headline of a 1958 *Life* article.[21]

Some popular accounts did note the need for caution. A 1949 article in *Harper's*, for example, warned about "the public misunderstanding generated by the few well-publicized but limited triumphs that have been achieved with the radioactive material and atomic rays."[22] Others tried to dampen inflated expectations about chemotherapy. And some noted that, as Amanda Dushkin quickly discovered in March 1960, the new chemical agents produced extremely unpleasant side effects, including nausea and vomiting.[23]

Nevertheless, the widespread faith in medical science undoubtedly helped to sustain Dorothy Dushkin's initial belief that the new drug administered to Amanda had wrought a miracle and that her terrible ordeal still might end happily. Other relatives as well as Smith College students also expressed confidence that Amanda would prevail. A letter from an Israeli cousin in June 1960 thanked Dorothy for her "reassuring word of Amanda's marked and miraculous cure!"[24] After the setback the following November, the cousin wrote again, this time stating that all Amanda's troubles soon would "prove a forgotten night-mare."[25] In a memoir of Smith College life in the late 1950s, Susan Allen Toth recalled that although Amanda "periodically disappeared" for treatments, "no one really believed she would lose her battle with cancer." Her classmates "thought comfortingly about the expertise of the world's best doctors at Mass. General." "It was impossible to imagine Amanda dying."[26]

Keeping Secrets

To some extent, growing optimism about cancer helped to engender a new openness. As two professors at the University of Minneapolis Medical School explained, patients could learn the truth because a cancer diagnosis no longer "carried the knell of death to its victim." As a result of 'the tremendous advances in diagnosis and treatment," the situation was "far from hopeless."[27]

Patients, too, became somewhat less secretive. Historian James T. Patterson notes that early-twentieth-century patients who learned the name of their malaise often refused to reveal it, even to intimates.[28] But Toth's memoir reminds us that Amanda Dushkin's classmates clearly knew she had cancer. Patients fortunate enough to regain health after finishing treatment were especially likely to disclose their diagnoses. Joining Cured Cancer Clubs, some proclaimed their identity as survivors. First established in 1939, those clubs gradually spread throughout the county.[29] According to a 1956 article in *Today's Health*, members were "former victims of cancer who have recovered and lived to carry out their vital idea: to provide a morale-building source of hope and reassurance for cancer patients by recounting their own fight against the disease, and to give financial aid to people impoverished by long sieges of cancer."[30]

Reach for Recovery provided similar services for breast cancer patients. In a 1954 magazine article entitled "I Had Breast Cancer," the founder, Terese Lasser, declared, "A deplorable curtain of silence hangs about this subject and it is time we lift it."[31] Lasser attempted to breach that silence by organizing a cadre of survivors to visit other women immediately after surgery. A far more famous survivor was the golf champion Babe Didrikson Zaharias, who underwent surgery for rectal cancer in April 1953.[32] A *Reader's Digest* article described her surprise visits to the hospital rooms of other patients: "The faces of anxiety-torn men and women turn hopefully towards her. She looks and is the picture of vibrant health. She discusses cancer with them as casually as she discusses her golf score. If they are facing an operation, she sits down and gives them words of encouragement from her own experience."[33]

But the growing openness about cancer did not unsettle the traditional evasion about mortality. One reason Zaharias could use her case to encourage others is that no one told her that her surgery had failed to remove all the cancer and that her prognosis therefore was grim. Her close friend Betty Dodd later explained why she and Zaharias's husband concurred with the doctors' suggestion that they withhold that information: "Neither one of us thought she could handle it—the fact that she wasn't going to live. Now, she could handle everything else. But she couldn't handle that. She never was told that—never, ever."[34] She died of the disease at the age of 45, three and a half years after her diagnosis.[35] Humphrey Bogart's son noted that his father similarly was not told that his esophageal cancer was far more extensive than he assumed in the fall of 1956: "In the movies my father had died many times, particularly in the early days. By 1942 he had made forty-five films. In them he was electrocuted or hanged eight times, and shot to death twelve times. But in reality, Bogie had an incredible will to live and he was nowhere near ready to die."[36]

And many patients never learned they had cancer. Despite the greater honesty Cured Cancer Clubs preached, studies suggested that the great majority of physicians still did not inform patients of cancer diagnoses.[37] Some physicians used euphemisms, speaking of a "mass" or "lesion," rather than a "neoplasm" or "cancer." A few asserted that a cancerous tumor was benign. When treatment was necessary, doctors gave just enough information to obtain compliance.[38]

Researchers explained their findings in various ways. One was that medical schools provided no instruction about how to deliver bad news. Another was that doctors continued to view cancer as an invariably fatal condition. Most had little faith in either the treatments or the early diagnosis which cancer advocates relentlessly touted. The prevalent attitude was "What's the use? You make an early diagnosis, the patient goes through a horrible operation and suffers, and two years later he's dead anyway." Concealment also reflected doctors' own discomfort with death; to them a fatal illness increasingly represented a "major defeat." In addition, doctors assumed that patients could not bear to anticipate their own deaths. "Quite representative," according to one researcher, "was the surgeon who stated, 'I would be afraid to tell and have the patient in a room with a window.'"[39]

Historians can offer other explanations as well. As we have seen, doctors had long concealed both prognoses and diagnoses. The enormous stature of postwar physicians facilitated evasiveness; patients were loath to question doctors or challenge their pronouncements. And secrecy was widespread throughout society. Parents of adopted children concealed the children's birth stories from them as well as from most outsiders.[40] Gays remained in the closet. Both scientists and government officials insisted that the cold war national security state demanded the classification of a widening circle of knowledge.[41] Doctors who suppressed painful news thus may have assumed they had support from the broader culture.

As in the past, physicians typically told at least one family member the information withheld from patients.[42] Nevertheless, family members, too, could have difficulty learning the whole truth, especially at the beginning of illness trajectories. Dorothy Dushkin's diary suggests that in their initial conversations with her, Amanda's doctors concealed the gravity of her daughter's disease and exaggerated therapeutic possibilities. In Peter De Vries's semi-autobiographical 1961 novel *The Blood of the Lamb*, a physician obfuscated the truth by furnishing a cancer diagnosis but then denying the disease inevitably would prove fatal. Had the father listened carefully, he might have heard the family doctor admit that the daughter's leukemia was incurable. But the doctor tried to drown out all doubts by expressing overwhelming confidence in modern medical science. "'My dear boy,' he began, 'where have you been the last ten years? There are first of all the steroids—cortisone and ACTH—which give a quick remission. The minute she's pulled back to normal with those, [the specialist] will switch her to the first of the long-range drugs, some of which he's helped develop himself. If they should wear off, there's—but let's cross those bridges when we come to them.'"[43] In his memoir of his son Johnny's death in 1947, John Gunther noted that he learned the deadly nature of the brain tumor only after he "peeped" at the medical records lying on a technician's desk: "There it was as clear as daylight—Johnny's tumor was 'undergoing glioblastomatous transformation.' That prefix 'glio'! No doctor had quite dared to tell us."[44]

Family members occasionally disagreed with physicians' decisions to shield patients from the truth. When social workers at New York's Presbyterian Hospital asked the adult son of a patient with advanced cancer to plan his "terminal care," the son wrote:

> A point that has been distressing me greatly is that my father does not know his true condition. I wonder if it is not a mistake of the hospital to give him the impression that he is on the road to recovery. Don't you think it will be a terrible physical and mental blow to him when he finds his true condition . . . ? Would it not be better to tell him that as a result of his operation most of his disease had been removed but some of it was difficult to get at, and that there is a possibility it may recur in the future? In this way, he would know he is not actually cured and could mentally prepare himself for the time of his gradual physical deterioration.[45]

Far more frequently, however, relatives actively participated in the culture of denial. According to another Presbyterian Hospital social worker, a man with metastatic stomach cancer seemed "more relaxed" after an operation although still in pain because he believed he was "on the way to recovery." He was about to leave the hospital under the care of his wife, who had told him that he had a "slow-healing ulcer" and would "require a very long convalescence."[46] A third social worker reported that the wife of a terminally ill patient was "very concerned that he should never find out the truth and kept saying that this was what the doctors particularly recommended. She is determined to abide by this although she finds it very difficult. She said that the doctors had advised her not to tell friends or too many relatives in case [the] patient should find out anything from them."[47] Stephen Bogart wrote that when his father was dying, his mother, Lauren Bacall, "sternly warned anyone who wanted to visit that if they were going to fall apart they should not come. She insisted that everybody be upbeat. This was not a death watch."[48] Writer Laura Furman recalled the entire family "enacting a charade" throughout her mother's terminal illness in 1959.[49]

Unsurprisingly, the silence surrounding children's cancer was especially profound. In DeVries's *Blood of the Lamb*, the father tries to keep his daughter from discovering she has leukemia through "the ceaseless censorship of work and tone" and "the hoarding of our secret from friends and neighbors."[50] A haunting 1962 *New Yorker* story by DeVries's wife, Katinka Loeser, describes a mother who continues to speak to her daughter in a matter-of-fact, occasionally playful manner, even after learning that the disease had advanced to the point where the girl could "slip away" at any time.[51] Although Amanda Dushkin was nearly an adult, her mother, Dorothy, considered "the hardest thing to bear" was "the idea of her realizing her own state of precarious life—of facing death." When Amanda died without having asked her parents "if her illness was fatal," Dorothy expressed relief. "All our dread & apprehension at having to answer such a demand was needless."[52]

Despite the evasiveness around terminal illness and death, many patients were aware they suffered from life-threatening conditions. Some doctors refused to lie. Some patients eventually convinced either a medical provider or a family member to speak honestly. And some discovered the truth by accident, such as through the name of a hospital or the type of treatment administered. Babe Ruth guessed his diagnosis when he entered New York's Memorial Hospital for Cancer and Allied Diseases.[53] Photo displays in leading popular magazines of the new machines used to treat cancer also made secrecy difficult. John Gunther recalled that the first time he saw his son Johnny "really frightened" was when he got ready for the first X-ray treatment. "He said to me again and again, anxiously, 'It's just for pictures, isn't it?' Then he knew from the time he spent under the machine that something much more serious than taking pictures was going on, and that this must be a form of treatment. He turned to me firmly and asked, 'Does this mean that I have cancer?' Then he murmured to [his mother] later, 'I have so much to do! And there's so little time!'"[54]

Patients often knew experientially when death drew near. "It is very difficult to fool a cancer patient very long," conceded one physician. "The patient knows intuitively that the situation is grave when he progressively loses weight and becomes weaker."[55] Patients may have been especially likely to recognize the truth if the deaths of relatives had rehearsed their own. Recalling the family's silence about her mother's approaching death from ovarian cancer, Laura Furman pondered: "How she could have believed that she would live, feeling as sick as she must have felt, I cannot imagine, especially given the fact that her mother had died of ovarian cancer fifteen years before."[56] A strong possibility is that Laura's mother tried to hide her awareness to protect her husband and children. As a 1950 Chicago study concluded, terminal cancer patients often responded "gallantly" to family efforts at concealment, "avoiding any mention of the facts as carefully as do those about them. Yet they know well what the real situation is."[57]

Redefining the "Good Death"

At a time when many Americans believed that medical science had a cure for virtually every affliction, popular culture no longer enjoined everyone to prepare for death. Most normative expectations for patients applied only to those deemed likely to recover. Thus, Talcott Parson's famous "sick role" theory, first enunciated in 1951, described people who were temporarily ill and soon would resume normal social duties.[58]

Nevertheless, we can discern some notion of appropriate behavior when facing mortality. Although fortitude remained a primary virtue, the spiritual elements of the *ars moriendi* tradition had largely withered. Privately, of course, many patients continued to try to make peace with finitude and prepare for eternity. The public discourse, however, insisted that patients marshal all available medical resources for the fight for survival, continue to hope for a cure, and remain engaged in ordinary life as long as possible.

Hospital personnel represented an important source of instruction about how to die. According to a 1965 sociological study, medical staff expected the patient to "maintain relative composure and cheerfulness. At the very least, he should face death with dignity. He should not cut himself off from the world, turning his back upon the living; instead he should continue to be a good family member, and be 'nice' to other patients. . . . He should cooperate with the staff members who care for him, and if possible he should avoid distressing or embarrassing them." Many nurses furnished cues to dying patients about appropriate behavior; a few scolded or threatened those who failed to conform.[59]

Margaret Harding illustrates the kind of patient who won approval.[60] According to New York's Presbyterian Hospital Social Service records of 1944, she was "75 years old, Irish by birth and parentage and Protestant in religion." The social worker noted that Margaret was "a most cooperative person and very uncomplaining. She talks very little about her condition and does not seem outwardly to be fretting over this; it is questionable whether she realizes the probably seriousness of this." The social worker saw Margaret again after her discharge when she went to the radiotherapy clinic for palliative treatment. There, she was reported as "cheerful, taking a motherly interest in other patients." Writing to request a placement for Margaret in a home for incurables in June 1945, the social worker stressed, "She is a person of real fortitude and understands what her condition is and that the outlook is not good. She has accepted this with a good deal of courage and is most cooperative. . . . We hope very much that you will be able to accept Miss Harding, as she is a person who will fully appreciate the good care you are able to offer patients."[61]

Although Cured Cancer Clubs encouraged most patients to assume they would rapidly return to health, members occasionally encountered people with no hope of recovery. The clubs had lessons for them too. According to a 1953 magazine article, Priscilla Dexter Kern, the president of the Washington, D.C., chapter, received a phone call from a "distraught man" whose wife Harriet had incurable liver cancer and "had refused to move from her bed for the last three months, stubbornly awaiting death." Two weeks later the man called again, this time to announce a "miracle." After receiving letters and cards from club members, his wife "got out of bed . . . for the first time in four months, called up her sister and went downtown to lunch with her." Despite the hopes of her many well wishers, Harriet "died several weeks later." Nevertheless, "that man will always remember his wife as the happy woman he loved instead of an embittered invalid."[62]

A 1956 article described Al Gilbert, a lung cancer patient, who had decided to kill himself after learning he had less than a month to live. In response to a phone call from his wife, Kern told him about members who had conquered cancer as well as about "more unfortunate acquaintances who had died of cancer, but who had met their deaths cheerfully and courageously." Although Kern was "a devout Episcopalian" and came from "a long clerical background," she said little "about a higher life after death." Secular encouragement, however, proved sufficient. By the end of the call, Gilbert had decided not to commit suicide. "Of course Al Gilbert died shortly afterward. But he met death cheerfully and with faith" instead of as a "depressed, embittered man."[63]

In the postwar period, as throughout the nineteenth century, both the personal and public writings of parents described dying children who had fulfilled the contemporary ideal. In February 1961, Dorothy Dushkin wrote that Amanda's "spirit and courage are wonderful to see." There was "never a whimper from her, no giving in to self-pity or running for comfort. An infinitely more admirable spirit than mine."[64] John Gunther's widely published account of his son's death helped not only to memorialize the boy but also to define the good death for others. In the foreword, Gunther noted that he wrote "the story of a long, courageous struggle between a child and Death" with the hope that other "afflicted" children and their parents "may derive some modicum of succor from the unflinching fortitude and detachment with which [Johnny] rode through his ordeal to the end."[65]

Although both Amanda Dushkin and Johnny Gunther had at least some awareness that they faced premature death, they continued to look forward to the future. Two days before Amanda's death, Dorothy wrote that "the long battle" against cancer "did not defeat her spirit. Never did she indicate doubt that she would come back into college life and take up all the activities she so loved."[66] When Johnny Gunther learned he could not finish his senior year at Deerfield Academy, he received help from a tutor and was able to graduate with his class. In his application to Harvard he wrote, "I wish to go to college primarily to complete a sound general education and to prepare myself for the years to come. Also, I wish to prepare myself for research work in physical chemistry."[67]

The psychoanalytic insights that spread widely during the mid-twentieth century meant that parents had somewhat more leeway than parents a century earlier to acknowledge children's deviations from the ideal. William Menninger, a psychiatrist and founder of the Menninger Clinic, and Franz Alexander, a psychoanalyst who fled Hitler's Germany, gave a uniquely American twist to Freud's theory, which then, in the words of a political scientist, "took the country by storm."[68] Dorothy Dushkin and John Gunther invoked similar dynamics to explain rare instances of unpleasant behavior. "Of course [Johnny] was fretful sometimes," Gunther confessed. "Sometimes he was subconsciously hostile to me as if out of resentment of my good health."[69] A month after Amanda's death, Dushkin ruminated on their interactions: Amanda's "acute illness & her necessity to fight it made her often very irritable with us. In a subconscious effort to hold on to life she fought those who were close to her as if we were to blame. It was a measure of her extremity that she, of such affection normally, should turn upon her parents with sudden accusations over little bedside matters."[70] In general, however, the children's good humor prevailed.

Celebrity Deaths

Narratives of celebrities' deaths helped to publicize the new model of the good death to a wide audience.[71] As silence about cancer gradually waned, newspapers and magazines increasingly disclosed the diagnoses of public figures. Although initial accounts of Babe Didrikson Zaharias's cancer presented her as triumphing over her disease, the many reporters who followed her career eventually had to acknowledge her grim prognosis and trace her gradual decline. Accounts of the illness of secretary of state John Foster Dulles followed a similar trajectory—first celebrating his recovery from an initial bout of gastrointestinal cancer in 1956, then expressing confidence in his ability to surmount his recurrence in February 1959, and finally recounting his deterioration and death later that spring.[72] Senator Robert Taft took pains to ensure that his diagnosis of metastatic disease in 1955 remained a carefully guarded secret. His obituaries, however, divulged the cause of death. Soon afterwards, "Eight Weeks to Live: The Last Chapter in the Life of Senator Robert A. Taft," by Jhan Robbins and June Robbins, appeared in *This Week* magazine, followed by a reprint in *Reader's Digest* and then publication in book form.[73]

According to widely circulating accounts, all three celebrities exhibited composure, even when learning the worst. Just as Dulles "took the news" of metastasized cancer calmly, so Taft received a diagnosis of terminal disease "without the flicker of an eye."[74] Although Zaharias received only partial information, she, too, met unpleasant news with equanimity. When "told she had developed another cancer," she "didn't flinch."[75] But acceptance and resignation no longer were central values. The *New York Times* obituary of Zaharias stressed that she "had fought valiantly against cancer for the last several months. She remained confident almost to the end that she would get well."[76] After announcing the "shock" of Dulles's cancer in February 1959, an editorial in the *Times* promised, "Mr. Dulles has great courage and great endurance. If any man in the condition indicated can recover, Mr. Dulles can."[77]

All three individuals also strove to remain involved in the activities that had made them famous. "Mrs. Zaharias Expected to Return to Golf Play," read a *New York Times* headline a month after her second cancer surgery.[78] Dulles began to experience the symptoms of his cancer recurrence shortly after Khrushchev issued an ultimatum to the Western powers demanding that they leave Berlin. A biography by Deane Heller and David Heller published the year after Dulles's death described him as "magnificent" in "meeting the Russian challenge," in terms of both "the strategy he devised" and his "devotion to duty." Although Dulles was dying, he "grimly rallied the free world to the defense of Berlin so effectively that the possibility of yielding to the Russian threats was never even seriously discussed." At times, his "pain grew so intense that he was hardly able to stand, yet when he conferred with foreign diplomats, his posture was ramrod erect, his voice strong and firm. The President, Dulles' doctors, members of his family, friends, devoted members of his staff, all implored the Secretary of State to 'take it easy,' to stop driving himself, to get more rest." Dulles, however, resisted all such appeals.[79]

Taft, too, had a job to complete before dying. As Jhan and June Robbins wrote, "he earnestly felt that the success or failure of the Republican administration's first year in office rested on his ability to get the congressional decks cleared" in time for Eisenhower to present his new agenda in the fall. Taft therefore continued to go to the Senate. Harry Truman later remarked, "You'd never have thought he was sick. Bob Taft was a man with a lot of guts!"[80]

The celebrities also remained cheerful throughout their ordeals. None was reported to express bitterness, anger, self pity, or despair. Four days before her death, Zaharias was "in good spirits," although "losing ground in strength."[81] Dulles and Taft continued to perform acts of kindness. "Even in the midst of his fatal illness," Deane and David Heller wrote, "the Secretary of State still found time for the considerate gestures which had become a Dulles trademark."[82] One of the last letters he dictated was to his London tailors, thanking them for altering his shirts.[83] Taft roused himself from a state of semiconsciousness to bid his wife farewell. "The visit lasted fifteen minutes," his biographers wrote. "Martha Taft's last glimpse of her husband showed him sitting up, waving cheerfully and smiling his famous win-or-lose campaign grin." He lapsed into a coma a few minutes later and died the following morning.[84]

We saw that throughout the nineteenth century all deaths—and especially those of famous people—served as reminders that the end could come at any time and that one must always be prepared. The cancer deaths of celebrities in the 1950s taught very different lessons. Rather than raising spiritual or existential issues, they underlined the urgency of redoubling efforts to conquer the disease. Zaharias established her own fund to help researchers cure cancer, pointing to her experience to inspire the public to donate.[85]

Other cancer advocates used the fatal illnesses of Dulles and Taft to raise money. A week after the announcement that Dulles's cancer had returned, Dr. Howard A. Rusk wrote an article in the *New York Times*, explaining not only the nature of Dulles's disease and the type of therapy his doctors would administer but also the reasons for optimism. The mortality rate of cancer of the digestive tract already had declined as a result of "advances made in surgery, the most effective weapon in the war against cancer." Now congressional committees were holding hearings on legislation that would vastly increase support "of an all-out international war against cancer" as well as other terrible afflictions. "Today Secretary Dulles fights against time for his life," Rusk concluded.

> When will the scientific break-through come to solve the riddle of cancer? Tomorrow, or in the indefinite future? No one knows. What we do know, however, is that the more scientists who are at work on the problem in laboratories all over the world the greater are the odds for solution and the quicker that solution will come. What greater evidence of appreciation to Secretary Dulles for his dedicated service to our country and world peace could the nation give than to add to its national effort against cancer an international attack? Cancer sufferers throughout the world would be given new hope in the realization that an international army of scientists was fighting for them. For some or for many victory might come in time.[86]

The day after Taft died, a senator asked Congress to set aside $100,000 of National Science Foundation funds for "Robert A Taft Cancer research scholarships."[87] And when President Nixon decided to support the National Cancer Bill of 1971, which vastly expanded federal funding for cancer research, he cited the deaths of Dulles and Taft as examples of the kinds of needless tragedies that must be averted.[88]

Paying the Bills

In one way Dorothy and David Dushkin were fortunate. The cost of health care had risen dramatically since the early twentieth century, when sickness had sparked more worries about wage loss than about paying for doctors and hospitals. By the early 1960s, medical expenses were beyond the reach of many Americans.[89] Chapter 6 notes that the consequences for many terminal cancer patients were especially serious. The Dushkins, however, never doubted that they could provide the best physician services and hospital care for Amanda.

Dorothy's diary partially explains how they covered the cost. Soon after Amanda Dushkin's first hospitalization, Dorothy wrote that David recently had made and sold some musical instruments. "It will be nice to have a little money coming in," she commented. "Expenses were draining us, even before A's illness, & now we have paid a $730 hospital bill & are waiting for bill from surgeon & dr. We shall weather it, with selling some more of Aunt A's stock & the last of bond came due." The parents also would economize, asking an older daughter to end her psychoanalysis and foregoing improvements at their summer camp.[90]

Although they managed to cope with the cost of Amanda's care during the following two years, the Dushkins appear to have lacked a major source of support on which many other Americans could draw. Hospital insurance plans spread widely during the postwar period. As the next chapter demonstrates, that growth helped to fuel an enormous expansion of hospital care, shifting the site of death and dying.

CHAPTER 6

"Nothing More to Do"

Death and dying increasingly moved into hospitals during the decades immediately following World War II. By 1960, 50 percent of deaths occurred in those facilities, and many patients who died elsewhere spent time in hospitals during the last year of life.[1] A key reason was that hospital care expanded rapidly after the war. The 1946 Hill-Burton Act provided federal grants to states to construct community hospitals. Within twenty-five years, the program disbursed $2.5 billion, adding 350,000 beds to the nation's hospital system.[2] The requirement that Hill-Burton hospitals provide free care combined with the spread of private insurance plans enabled patients to fill the beds. Largely because government policies encouraged large corporations to offer health insurance as a fringe benefit to union members, the proportion of the civilian population with hospital coverage rose from 30 percent in 1946 to 45 percent in 1950 and then to a stunning 74 percent in 1961. Between 1950 and 1960, hospital admissions increased nearly forty percent.[3]

But dying patients were not hospitalized solely because those institutions grew in size and number. Rather than seeking to lower their death rates, some hospitals now wanted dying patients who could serve as research subjects. And the patient population was much sicker than it had been before the war. Although the two most common causes of hospital admission in the first half of 1960 were "childbirth without complications" and "hypertrophy of tonsils and adenoids," a growing number of patients were seriously ill. Many underwent extremely complex operations, including heart valve replacements and major cancer resections.[4] Many others arrived in the midst of acute vascular incidents. Rising expectations of cure encouraged these people to seek hospital placement; a high proportion met death instead.[5]

Few nurses were available to care for the greatly expanded and very sick clientele. Since the late nineteenth century, hospitals had relied on student nurses and the private-duty nurses hired by wealthy families; beginning in the 1940s, however, administrators employed more and more nursing-school graduates. To save money on a service they had previously received for free, facilities kept staffing levels low.[6] Margaret Nestor, a New York City nurse, described the difficulty of caring for many extremely ill patients in October 1942 letters to her husband, a soldier stationed in California. "Just gone home from work," she wrote late one evening, "was so busy that I wasn't able to get off one of my patients went bad about 4:45 PM & had a plenty tough time to pull him through. . . . Had all my work to do after 7:00 PM that I should have gotten done between 5:00 & 7:00." Among her many cases a few days later were three patients with pneumonia ("one of them a real sick girl 28 years old"), three with coronary problems (one of whom was "getting oxygen constantly, not even allowed to move his self so we have it all to do"), and several with cancer (including one who probably would "kick out in a day or so" and "a Leukemia that got another blood transfusion to-day & had a chill following it so that was a big job").[7]

The departure of many nurses for the front helps to explain why civilian nurses like Nestor had more responsibilities than they could handle. But the workload continued to be overwhelming even after the war, when a severe nursing shortage forced hospitals to remain understaffed at the same time that the level of patient acuity rose. In 1959, Lenox Hill Hospital in New York employed 90 professional nurses, 56 fewer than were needed.[8] In many hospitals, one nurse was responsible for as many as 60 patients during the day and 200 at night.[9] And a growing number of these individuals required complex care. A study by Cleveland's University Hospital found that between 1938 and 1953, nurses cared for 900 percent more patients using oxygen therapy and 177 percent more patients with intravenous lines.[10]

Architectural changes compounded the difficulty of tending large numbers of extremely ill, labor-intensive patients. Nurses traditionally had monitored the sickest by placing them close to the central station in the ward. As private and semiprivate rooms gradually replaced large, open wards, supervision became more difficult.[11] When an elderly woman with a heart problem in Kingston, New York, developed pneumonia in 1952, her physician sent her to the local hospital to receive oxygen. Fearing that the illness would overtax her heart, he insisted that she not be placed in a private or semiprivate room unless she could afford to hire a private-duty nurse. Only in a ward could she receive the "constant attention" she required.[12] By the early 1950s, however, many employer-funded private insurance plans provided reimbursement for semiprivate rooms; as a result, patients often refused to enter wards, regardless of the severity of their conditions.[13]

In a situation where a small number of nurses tried to save many desperately ill people isolated in separate rooms, bestowing sympathy and compassion on those who were slowly dying assumed a very low priority. A 1959 *New York Times* article about nursing at Lenox Hill Hospital observed that "something the nurse has little time to give the patient is TLC—tender loving care. . . . Sometimes tender loving care can be an important adjunct of medical attention, but it is rare when a nurse these days can . . . pay the role of the angel in white." The reporter described "Mrs. Slattery," an 81-year-old woman, "alone in life," who lay on her side softly moaning. She was "too frail to move, and her wasted fingers [clung] to the bed's side bars." The nurse changed the bed, turned Mrs. Slattery to her other side, and requested more pain medication when her moans interrupted another patient's birthday party. But the nurse could spare only "a few minutes" to talk to the woman, who remarked that "she has just asked the good Lord to take her away."[14]

Intensive Care Units

The primary response of hospitals to both the nursing shortage and the elimination of open wards was to establish intensive care units (ICUs), which gathered critically ill patients together in one place and assigned special nurses to watch them full time. (Although ICUs today tend to be associated with the most complex life-extending technologies, the earliest units rarely contained more equipment than was available elsewhere in the hospital.) Six years after North Carolina Memorial Hospital in Chapel Hill established the first ICU in 1953, a survey found 238 units in short-term, private nonprofit hospitals. By 1965, the number had grown to 1,040.[15] As historians Julie Fairman and Joan Lynaugh conclude: "Critical care accentuates and enables a tendency long noted in American health care ideology—the supremacy of curing disease and prolonging life over all other considerations. The invention of and massive investments in critical care and critical care nursing was, one could argue, simply a manifestation of the power of that ideology. Even now, critical care remains the most expensive and fastest growing type of care offered in hospitals."[16]

Admissions criteria for the ICU excluded dying patients. According to an article on the seventeen-bed unit at Connecticut's Manchester Memorial Hospital, the patients "might be described as 'super-critical' but not, [the administrator] emphasizes, moribund. He wants it clearly understood that 'special care' is not to be equated with 'terminal care.'"[17] Baltimore City Hospitals made the same decision. "Due to limitations of bed space in the unit," two physicians noted, "it was found to be more important to guard against the admission of non-salvageable patients."[18] An article by both an administrator and a surgeon at Mary Hitchcock Memorial Hospital in Hanover, New Hampshire, explained the screening process somewhat differently: "To maintain good morale among the nurses working in a special care unit there must be some possibility of a successful outcome to crown their strenuous labors. Loading the unit with terminal cases is the best way to extinguish the optimistic spirit and drive so essential for good nursing care and good morale in the special care unit." The authors added that patients would be more willing to enter the ICU when they understood that its purpose was "to extol the salvage rather than mark the failures."[19] Patients who failed to demonstrate sufficient improvement within a specified period were returned to general hospital units.[20]

But it was not always possible to predict who would be a "salvage" and who one of the "failures," or to discharge everyone who did not recover quickly enough. Although studies repeatedly found that ICUs saved lives,[21] death stalked the new units. Community Hospital of Battle Creek, Michigan, reported that in 1960 its ICU had a "large number of deaths—58, or 15.2 percent of all discharges from the Unit." More than one-fourth of all hospital deaths took place in that unit. The report stressed that "neither staff nor patients" regarded the ICU as a "terminal care facility"; nevertheless, the death ratio "considerably exceed[ed] predictions."[22] The authors of the article on Mary Hitchcock Hospital similarly acknowledged that its ICU had a high mortality rate.[23] Fairman and Lynaugh note that ICU nurses throughout the country "were confronted with an unrelenting stream of physiologically fragile, disfigured, often comatose patients and the daily occurrence of death."[24] ICUs thus faced in an exaggerated form the problem encountered by the larger institutions—attempts to exclude the moribund could not eliminate the presence of death.

Studying Death and Dying

The late 1950s and early 1960s witnessed not only the emergence of intensive care units but also some of the first stirrings of a movement to restore dignity to the dying. Writings by both physicians and lay people questioned the practice of concealing cancer diagnoses and poor prognoses.[25] Others challenged the overriding emphasis on prolonging life. A widely discussed anonymous essay in the January 1957 *Atlantic Monthly* began this way: "There is a new way of dying today. It is the slow passage via modern medicine. . . . If you are going to die it can prevent you from doing so for a very long time."[26] Two months later, *Reader's Digest* published a condensed version of the article, reaching a much larger audience.[27] Doctors, too, were impressed. The editors of the *New England Journal of Medicine* declared that the essay "should be required reading for physicians."[28] Writing in the *Journal of the American Medical Association*, Dr. Frank J. Ayd used nearly identical words when recommending the article to his colleagues.[29]

Herman Feifel's 1959 edited volume *The Meaning of Death*, which included essays by experts from various disciplines who challenged the cultural denial of death, also attracted considerable attention.[30] One reviewer pointed to the conspicuous absence of sociology,[31] but major studies in that field soon appeared. The first, *Boys in White: Student Culture in Medical School*, by Howard S. Becker and colleagues, touched only briefly on death.[32] The others, however, focused on patients at the end of life. Three books—Barney G. Glaser and Anselm L. Strauss's *Awareness of Dying* (1965) and *Time for Dying* (1968) and Jeanne C. Quint's *The Nurse and the Dying Patient* (1967)—as well as occasional articles by the same authors, were based on field work collected at San Francisco Bay Area hospitals between 1961 and 1964. The fourth, *Passing On*, by David Sudnow, relied on nine months of observation at a county hospital and five months at a private facility between 1963 and 1965.[33] Signaling a rising discontent with hospital care for the dying, these works also provide a unique window on that care. Together, they describe medical professionals who knew little about tending the dying, felt extremely uncomfortable doing so, and tried to concentrate on patients most likely to recover.

Boys in White argued that medical schools imbued students with a strong sense that "the true work of the physician" was "saving endangered lives." Snap quizzes, for example, commonly assumed that a life hung in the balance. When a student answered a question incorrectly, a staff member might say, "Come on now, Old Brown is lying there bleeding to death or something. We'd better find out what it is, don't you think? After all, there he is—he's your patient. He's going to die if you don't do something pretty quick, isn't he? But he doesn't have to die. You can save his life if you want to, if you know how to." That emphasis provided "a basis for classifying and evaluating patients; those patients who can be cured are better than those who cannot."[34]

Because much of the direct patient care was delivered by nurses, the other studies focused largely on them. Quint found that nursing schools, too, stressed the "life saving goals of medical practice."[35] Discussions about terminally ill patients tended to be confined to a few hours at the end of the course and to focus exclusively on care of the body immediately before or after death. Although psychological insights had begun to inform the curriculum, students learned little about communicating with people who were dying. Some never encountered such patients during their training. Virtually all left school convinced that care for the dying took time and effort that could better be spent helping others recover.

According to Glaser and Strauss, nurses' behavior after graduation reflected that bias. Just as nurses rarely chose to work in chronic care facilities, so they tried to avoid the "monotonous" task of caring for hospital patients who were slowly dying. Preferred assignments were in ICUs, emergency rooms, and premature baby units, where most patients either recovered or died quickly; the few who lingered in an indeterminate state or took too long to die were sent elsewhere.[36] Many nurses withdrew their efforts from patients with no hope of survival.[37] Glaser and Strauss refrained from commenting on the implications of the phrase "nothing more to do" but noted that physicians routinely used it to indicate that recovery no longer was possible and "only" nursing care was required.[38]

Some dying patients, however, received far more attention than others. In a 1964 article in the *American Journal of Nursing*, Glaser and Strauss reported that nurses bestowed the most care on patients with the highest "social value": "Extra 'good will' efforts are made to talk with them, to keep up their spirits, to make them comfortable, and to watch for sudden changes in their condition."[39] Age was a critical factor in calculations of patient worth. Because dying children were "cheated of life itself" and the deaths of people in their middle years often had a deleterious impact on families and society, nurses viewed losses from both groups as far more serious than those from the elderly, who had enjoyed long lives and no longer engaged in productive activities. (It also is possible that the dramatic drop in the childhood mortality rate and the increase in life expectancy meant that the death of people in their early years had begun to seem like an inversion of the natural order.) Other characteristics that influenced social evaluations included class background, "skin color," accomplishments, occupation, and education. In addition to making their own assessments, nurses observed patients during visiting hours to determine the extent to which their deaths would "matter" to family and friends. That analysis helps us understand why Amanda Dushkin, a Smith College student and the well-loved daughter of a prominent family, continued to receive good care even after treatment ended. "All nurses and doctors have done everything possible to make Amanda comfortable & cheered," Dorothy reported shortly before Amanda died.[40] Patients imbued with less social value, according to Glaser and Strauss, "tend to receive minimal routine care. In some few cases, [they] may receive less than routine care."[41]

Sudnow's description of death and dying at "County," the public hospital he studied, enables us to witness that care. Because all County patients were indigent and many were African American, physicians regarded them as "less than desirable social types."[42] The absence of private physicians meant that no staff members had special investment in individual patients. The relatively short average stay at County provided few opportunities for medical personnel to interact with patients. And, as we will see, families and friends rarely visited. Sudnow observed "a decided phasing out of attention given to 'dying' patients." Some were treated as if they were already dead. One nurse, for example, spent "two or three minutes" trying to close the eyelids of a woman who was still alive. "This involved slowly but somewhat forcefully pushing the two lids together to get them to adhere in a closed position." When asked what she had been doing, she explained that it was much more difficult to close eyes after death when the lids became "less pliable, more resistant, and [had] a tendency to move apart." As a result, she always tried to perform that task before patients died.[43]

Family Visits

"A general norm in American culture is that no one should die alone," observed Glaser and Strauss. "Preferably, death should be attended by a close relative."[44] Hospitals honored that norm by following the traditional practice of posting "danger lists," alerting the relatives of patients on the list and permitting those family members to remain beyond normal visiting hours. There was no guarantee, however, that the list contained the names of all patients at imminent risk of death.[45] Moreover, some hospitals continued to undermine their own policies.

Sudnow found that patients on the danger list at the county facility he studied "theoretically have the right to round-the-clock visitors," but "in actuality nurses strive to separate relatives from those patients whose deaths are regarded as imminent. They urge family members to go home and await further news there." Sudnow attributed that practice to the "matter-of-fact fashion" with which the staff handled the dying; both nurses and physicians felt free to disregard patients close to death only if families were absent. An additional explanation may have been that medical personnel assumed the families of the hospital's low-income, African American clientele could not remain composed when death occurred. Sudnow overheard one staff member remark, "These Negroes don't know how to control themselves." Regardless of the reason, the result was the same: It was "common for a patient to die unattended and be discovered as dead considerably later, when a nurse, aide or doctor happens into his room."[46]

As in the past, the ability of families to remain with patients in private hospitals depended on the type of rooms they occupied. Most hospitals instituted more rigid visiting hours for private rooms during the 1960s.[47] Nevertheless, some relatives were able to stay around the clock with dying patients. According to Robert V. Wells, the journalist Lewis Sebring remained all night "by his mother's side" when she died in a private room in a Schenectady, New York, hospital in 1953. He left briefly on an errand in the morning but soon returned. A cousin "arrived shortly thereafter and remained with Lewis for the last few hours."[48] Dorothy Dushkin felt entitled to hold an extremely extensive vigil. Shortly before her daughter

Amanda's death in a private room, Dorothy wrote that "with the exception of two brief interruptions," she had "been in daily bedside" for nearly two months. "Three weeks at Mass. General Hosp. in Boston—ten days at the College infirmary and now two weeks at Cooley-Dickinson Hosp. in Northampton."[49]

Other parents had very different experiences. Although most nineteenth-century hospitals had prohibited parents from visiting children, psychological theories about "separation anxiety" had led most institutions to encourage parental visits by the 1940s.[50] Nevertheless, parents continued to enjoy more freedom to visit offspring in private rooms than in wards. Before bringing her 6-year-old son Simon to Presbyterian Hospital in New York to receive palliative X-ray treatment for brain cancer, Mrs. Gerson had placed his name on the waiting list for a private room.[51] But none was available on the day of his admission, and she had been forced to accept a bed in a children's ward. A social worker reported that the day after Simon arrived in the ward, the mother appeared at the social service department "in tears and anger" because she had promised Simon "she would visit 3 times per week under [the] impression that those were the visiting hours." Although the superintendent's office previously had advised her to take Simon home "to give him love and attention," the office now refused to grant her a special pass so she could visit more frequently. The boy's "loneliness for her," she argued, would certainly aggravate his condition. Mrs. Gerson asked the worker to take a written note to Simon and read it to him. "She also asked if she might call daily to do this. She waited until worker returned from ward, asked numerous questions about patient, and left hospital somewhat calmer, though not less anxious." The worker noted that the mother "came to give messages and gifts to [the boy] on days she was not allowed to visit. She gave instructions about what to say to patient, how to handle him, how to hand toys to him, etc."[52] Although we can almost hear the worker's frustration with Mrs. Simon's frantic insistence on retaining contact and control, the mother's anguish and desperation are equally apparent.

A second Presbyterian Hospital record captures a dying child's desolation. Timothy O'Reilly, another 6-year-old boy with an inoperable brain tumor, cried for his parents for several days before his death.[53] The father of a 12-year-old girl who had died of Hodgkin's disease in another facility complained in a letter: "Even during the time our Sheila lay critically ill in a hospital ward, needing every bit of our help, my wife and I were censored by the supervisor for obstructing ward routine."[54]

Overworked nurses sometimes appreciated the personal care services relatives provided, especially during mealtimes and at night, when staffing levels were extremely low.[55] But as we have just seen, nurses also found that family members could exacerbate as well as lessen problems. "Had a terrible day," the New York City nurse

Margaret Nestor complained to her husband. A man had died late in the afternoon, and "it took three of us to hold him in bed until he took his last breath. Was getting oxygen. His wife, son & daughter were on the floor which made it worse with them around."[56] Nestor may have minded the ability of those family members to observe and assess the care she delivered at a particularly tense time.[57]

Another possibility is that they violated what Glaser and Strauss call the " 'sentimental order'—the intangible but very real patterning of mood and sentiment that characteristically exists on each ward."[58] Although Dorothy Dushkin complained about the "routine enforced cheerfulness" in hospitals, the long hours she spent by her daughter's bedside suggest she managed to display the appropriate affect. Other family members required instruction in proper behavior. As an article in *Time* magazine noted in 1964, "Relatives may need to be coached in deathside manners. If they have not already faced their own emotional problems, they may become depressed, or tearful or even hysterical."[59] Relatives who persisted in creating "scenes," by openly expressing rage, grief, or despair, frequently were encouraged to leave.[60] "Florida," Nancy Kline's short story about a young girl's visit to her dying grandfather in the hospital, also suggests that public hospitals were not the only ones that sought to remove families. The nurse insisted that the patient rest and urged the family to return in the morning. The grandfather died that night.[61]

Intensive care units in all types of hospitals severely restricted the presence of relatives and friends. "The problem of visitors is a tedious and difficult one," noted a 1954 article in *Modern Hospital.* "Visitors must be the exception rather than the rule. Regulations must be passed and rigidly enforced."[62] In general, ICUs permitted one visitor for a few minutes during one or two hours a day.[63] Although facilities occasionally allowed visitors to remain around the clock with terminally ill patients, most made few exceptions to the rules.[64] Elisabeth Kübler-Ross later described "Robert," a 21-year-old student with leukemia who died in a hospital ICU in the early 1960s "with his parents sitting outside in the waiting room."[65] Relatives of affluent patients thus began to experience what families of poor people had long endured. Regardless of family backgrounds, many ICU patients died alone.

It also is likely that even when family members received permission to stay with their dying relatives, hospitals subtly altered relationships. The growing use of technological treatments at the end of life made physical contact difficult if not impossible. When Johnny Gunther died, his mother "reached for him through the ugly, transparent, raincoat-like curtain of the oxygen machine." The story "Florida" describes the girl's first sight of her grandfather after his heart attack: "He has a plastic mask over his nose and mouth. Tubes come from underneath the sheets. Something is beeping." A North Carolina physician complained that family members "were barred from approaching the bed" of a dying man "by oxygen tank, suction apparatus, tubes for suction, catheterization and infusion."[66] Hospital rules further inhibited intimacy. A nurse reprimanded the girl in "Florida" when she tried to sit at the end of her grandfather's bed. And the "routine forced cheerfulness" Dorothy Dushkin decried could preclude the exchange of more authentic emotions.

Dying Patients as Research Subjects

Research conducted in hospitals during the decades immediately following World War II also reveals medicine's attitude toward severely ill and dying patients (especially those who were poor). Although investigators relied heavily on various vulnerable populations, including prisoners and the inmates of institutions for mentally retarded children, terminally ill patients were an especially critical group.[67] "Patient V is continuing downhill so you should be hearing from me soon," wrote a research assistant involved in a major experiment at Massachusetts General Hospital in the mid-1950s.[68] Despite the varied uses of the word "terminal," most patients with that designation were assumed to have limited life expectancy and thus were unlikely to experience the long-term effects of the most toxic agents. In some cases, investigators could correlate research findings with the results of autopsies.[69]

Researchers also drew overwhelmingly on poor patients. One investigator provided what President Clinton's 1994 Advisory Committee on Human Radiation Experiments called "a frank description of a quid pro quo rationale that was probably quite common in justifying the use of poor patients in medical research: 'We were taking care of them, and felt we had a right to get some return from them, since it wouldn't be in professional fees and since our taxes were paying for their hospital beds.'"[70] Some research subjects occupied free wards in private hospitals. In 1947, for example, Sloan-Kettering Institute for Cancer Research tested new chemotherapeutic agents on patients in ten beds "set aside in a special research ward at the Memorial Hospital." The study participants were "hospitalized, without charge to them, for any aspect of their care."[71] (It is unclear whether the patients or their families were aware of the reason the care was free.) Investigators found other research subjects in public hospitals. According to the Advisory Committee on Human Radiation Experiments, the use of such patients as research subjects reflected not only "a general societal insensitivity to questions of justice and equal treatment" but also the belief "that poor people were better off being patients at hospitals affiliated with research-oriented medical schools. . . . Such institutions, it was thought, offered poor people their best, and perhaps their only, chance to secure quality medical care."[72]

That view helps to explain the establishment of the two new 300-bed New York City cancer hospitals which opened in the summer of 1950. The Francis Delafield Hospital, located at 163rd Street and Washington Avenue, was closely affiliated with Columbia-Presbyterian Medical Center. Further downtown, at First Avenue and Sixty-seventh Street, the James Ewing Hospital was contiguous to Memorial Hospital and the Sloan-Kettering Institute and operated in conjunction with them. In both cases, the private institutions donated the land and appointed the medical staff. The city paid construction costs and assumed responsibility for administrative, clerical, nursing, and maintenance personnel.[73]

The two hospitals represented an enormous improvement over the New York Cancer Institute, the municipal facility on Welfare (previously Blackwell's) Island, now slated for closure. According to a widely distributed booklet about Ewing Hospital, visitors would easily note "the disappearance of the vast, old-fashioned hospital ward with 20 or 40 beds. In James Ewing, no ward holds more than six beds and some rooms have only one or two."[74] Publicity surrounding both Delafield and Ewing stressed that for the first time, city patients would have access to the best physicians and latest technologies. "The affiliation of the James Ewing Hospital with the Memorial Center and the Sloan-Kettering Institute will assure a high quality of medical care and extraordinary therapeutic facilities," declared Commissioner of Hospitals Marcus D. Kogel.[75] The *New York Times* emphasized the "several radioactive isotopes" that were "available to Delafield patients" as well as its 2,000-volt X-ray machine, enabling treatment of extremely "deep-seated" cancers.[76]

What city officials and newspapers failed to note was that a major goal of both hospitals was research. A year before Delafield opened, the secretary of the American Cancer Research Committee wrote that its wards would "be available for the cancer research bed studies."[77] Ewing's director, Cornelius P. Rhoads, later stressed that both hospitals were "originally built for research."[78] We last encountered Rhoads in 1949, when a *Time* magazine cover story featured him as an exemplar of the postwar cancer researcher. Then he had been director of both the Sloan Kettering Institute and Memorial Hospital. After Ewing's establishment the following year, that facility, too, came under his jurisdiction.

Neither Delafield nor Ewing completely fulfilled its research mission. In 1957, Rhoads reported that a sixty-bed general research ward at Ewing operated at 50 percent capacity and a forty-two-bed chemotherapy ward at 75 percent.[79] Although Rhoads did not provide figures for Delafield, he implied that its record also disappointed. One cause, he asserted, was the scarcity of competent nurses. The postwar nursing shortage was especially severe in public hospitals, where workloads were higher and salary levels lower than in private ones.[80] In 1950, the New York

Department of Hospitals acknowledged that an "excessive numbers of patients" were in its "crowded facilities with inadequate nursing personnel both in terms of numbers and skills."[81] The situation had not improved two years later, when the *New York Times* reported that the monthly pay for a nurse in a city hospital was considerably beneath that for a nurse at a private facility. As a result, many municipal hospitals operated with only 53 percent of the registered nurses required.[82] And Rhoads pointed out that "patients with advanced and painful forms of cancer who are under research study" required special care. "Nurses of outstanding character, patience, and competence must be available for the work, far more arduous and demanding as it is than ordinary hospital nursing duty."[83] The American Cancer Society augmented the salaries of nurses at both Delafield and Ewing, but complaints continued.[84] In 1956, Rhoads wrote to the commissioner of health that the "care given [Ewing] patients is not good." (An earlier version of the letter read, "The care is so poor as to be almost a scandal.")[85]

If the number of research subjects fell short of initial expectations, however, significant numbers of patients served in that capacity. In its biennial report for July 1, 1955, to June 30, 1957, the Sloan-Kettering Institute wrote, "From its inception, and throughout the development of the program, the Sloan-Kettering Institute research has extended from laboratory bench to bedside." The "clinical cancer therapy investigation is pursued by the research physicians of the Sloan-Kettering Institute functioning also as members of the medical staffs of the two associated hospitals. Much of this work is in the special research facilities provided by the James Ewing Hospital and the City of New York. The research wards serve as testing and training centers, perform pioneer work in developing better methods of treatment by studies in man, and train medical specialists to carry forward cancer chemotherapy in other institutions."[86] More succinctly, the commissioner of hospitals reported that Ewing patients "are the focal point of the Sloan-Kettering Institute laboratory program of cancer research."[87]

The 1949 *Time* cover story on Rhoads described the soul searching that Sloan-Kettering and Memorial doctors engaged in before administering experimental agents:

> In each individual case, the doctors have to make a grim decision. Should they prolong a life that is sure to be "unsatisfactory?" Should they, by prolonging life, place a crushing burden on the patient's family? Should they, in desperate cases when everything else has been tried, use a drug so dangerous that it may kill the patient immediately? . . . The doctors decide each case separately, considering

> such matters as the painfulness of the treatment and the patient's chance for happiness during his possible remission.
>
> Some cancer doctors admit that they have almost cracked up thinking about such things, and about their utter helplessness in hundreds of cases. Dr. Rhoads, too, has his moments of depression.[88]

An important piece of Rhoads's personal history, omitted from the laudatory *Time* article, suggests that he may have felt very differently about patients from different social groups. While conducting research in Puerto Rico in November 1931 under the aegis of the Rockefeller Institute, Rhoads had written: "Porto Ricans are beyond doubt the dirtiest, laziest, most degenerate and thievish race of men ever inhabiting this sphere. What the island needs is not public health work but a tidal wave or something to totally exterminate the population. I have done my best to further the process of extermination by killing off 8."[89]

Although Rhoads insisted that the comments appeared in a letter he intended to be both confidential and humorous, the leader of the Puerto Rican Nationalist Party, who found a copy, was not amused, and he publicized it widely. Investigations conducted by both the Puerto Rican governor and the Rockefeller Institute found no evidence of any "extermination." As a result, when a Rockefeller public relations expert helped Rhoads frame his version of the event, the furor gradually evaporated, and his career continued on its upward trajectory virtually unscathed.[90] But then, in 2002, a University of Puerto Rico biology professor discovered the letter and sent a copy to the American Association of Cancer Research, demanding that the association rename its Cornelius P. Rhoads Memorial Award presented annually to a young cancer researcher since 1979. After commissioning an independent investigation by a preeminent authority on medical ethics and human experimentation, the association stripped Rhoads's name from the award.[91]

We can only speculate about the extent to which physicians at Delafield and Ewing obtained consent from patients. By 1950, both the War Crimes Tribunal at Nuremberg and the American Medical Association had endorsed the principle of "informed consent."[92] The standards, however, were much lower for sick participants than for healthy ones.[93] Announcing the establishment of a special chemotherapy research ward at Ewing in 1952, both Rhoads and the commissioner of hospitals promised that all clinical trials "would be conducted on a voluntary basis."[94] But "voluntary" may have had a very restricted meaning in that setting. A professor of medicine at Yale during the 1950s and 1960s recalled that it was "very easy to talk a terminal patient into taking that medication or to try that compound or whatever the substance is. . . . It's very easy when you have a dying patient to say, 'Look, you're going to die. Why don't you let me try this substance on you?' I don't think if they have informed consent or not it makes much difference at that point." A physician involved in medical experiments during the 1940s later told interviewers: "One of the real ludicrous aspects of talking about a prisoner being a captive, and therefore needing more protection than others, is, there's nobody more captive than a sick patient. You've got pain. You feel awful. You've got this one person who's going to help you. You do anything he says. You're a captive. You can't especially if you're sick and dying, discharge the doctor and get another one without a great deal of trauma and possible loss of lifesaving measures."[95] Because public hospital patients had extremely few alternatives for care, they must have felt especially vulnerable.

Of course, understanding the research culture of the 1950s and 1960s also might soften any condemnation of Rhaods and New York City health officials. Historian Gerald Kutcher criticizes present-day studies that apply universalistic ethical principles, ignoring the standards in different periods.[96] And there is little information about how the researchers calculated the costs and benefits to the study participants. Did the investigators fear that the experiments might hasten death and inflict additional pain and suffering on cancer patients at the end of life? Alternatively, did the researchers view themselves as offering desperately sick people a last chance for survival? By the 1980s, many advanced cancer patients would press for access to clinical trials of experimental treatments, which often offered their only hope.[97] From that perspective, Delafield and Ewing patients might appear to have been uniquely privileged. Is it possible that the influx of highly competent nurses into the wards of both Delafield and Ewing helped to compensate for any harm the studies caused?

If we cannot draw any firm conclusions about the ethical issues involved, this story does illustrate two major themes of this book. One is the low value attached to the lives of people who were not only terminally ill but also very poor. In the past, New York City had banished that population to dismal island facilities bereft of the most rudimentary care. Now the city secured their admission to state-of-the-art hospitals, but only at a price.

The other is that medicine's duty to forestall death through research gradually eclipsed the duty to relieve the suffering of the dying. The history of Delafield and Ewing offers an extreme example of that hierarchy of concern. Administrators at both hospitals tried to limit admission to patients who could advance the research mission. The individuals should have "far advanced" cancers that had not responded to conventional treatments but still have at least some possibility of recovery as well as the strength to withstand arduous experimental procedures.[98] An early plan for the establishment of the hospital noted that its primary function was not to provide nursing care.[99] Reporting that a third of Delafield's beds remained empty in 1954, the *New York Herald Tribune* stressed that "it wasn't the hospital's desire only to get more patients but the right kind of patients." The chief of medicine was quoted as complaining, "We are getting too many patients who have been sent here as a last resort."[100]

Timely discharge also was critical. A 1957 study found that "considerable numbers of patients seem to remain in Delafield longer than a surgical or other acute treatment would require." Half of the 196 patients discharged between May and June 1957 had been "in the hospital for longer than a month; 54 were discharged after a hospital stay of 31 to 60 days, 24 after a hospital stay of 61 to 90 days, and as many as 20 had been in the hospital more than three months. Of these 20 long-stay patients, 10 died in the hospital." "In other words," the report emphasized, "these 10 patients remained in the specialized cancer hospital for protracted terminal care."[101]

Noting that the problem extended to Ewing as well as Delafield, Rhoads complained to the commissioner of hospitals that the two facilities were "not now being employed for serious research, as intended, because they are largely filled with terminal patients." Stressing that "terminal care is not their function," he urged that the department of hospitals to send to Ewing only patients "qualified" for "study" and "transfer" them "out when the study period is completed."[102] In his view, research clearly trumped care of the dying. Both Delafield and Ewing, however, faced the same dilemma as tuberculosis sanatoriums had confronted earlier in the century. Because few alternative facilities were available, institutions established to foster recovery and demonstrate the benefits of new therapeutic approaches found themselves caring for many patients at the end of life.

CHAPTER 7

A Place to Die

In the summer of 2000, archivists and a medical historian at Columbia University in New York rescued hundreds of Presbyterian Hospital patient files from the millions contained in a large warehouse slated for destruction. Searching through the salvaged files, I found forty-nine discussing cancer patients who had been referred to the hospital's social service department between 1940 and 1965 to make arrangements for terminal care either at home or in alternative institutions.[1] Most of the patients were very poor. Four were African American or Afro-Caribbean, and many others were recent immigrants from Europe or Russia. Thirty-five were men or, in two cases, boys. Although the number of these files is limited and they may not be representative, they provide rare insight into the experiences of an extremely vulnerable patient population that hospitals did not want. Very few sources of care were available.

Two contemporaneous reports suggest that the problems those patients encountered were not unique. A 1945 study by four social workers of 200 patients with terminal disease in Boston cancer clinics concluded that institutional services were so sparse that "there is little left for the dying patient except to remain at home and to get along as well as he can." Although public health nurses provided some free care, their visits were inadequate for the many people receiving injections to relieve pain.[2]

Five years later, the Institute of Medicine of Chicago published a far more extensive study, *Terminal Care for Cancer Patients*, that surveyed facilities and services in the Chicago area. The report gave a nod to postwar medicine's curative and research focus. It quoted a recent article in the *Journal of the American Medical Association* asserting that "not infrequently treatment intended to be palliative may be curative. The condition of no patient is positively hopeless as long as he is alive."[3] The study also argued that a greater emphasis on cancer's final stage could advance research interests. But the authors stressed that patients with terminal cancer deserved attention from "anyone with sincere concern for sick and helpless human beings." And despite its dry, dispassionate tone, the report repeatedly used the word "tragic" to refer to the plight of that group.[4]

The study defined the final stage of cancer as beginning when physicians recognized there was no "reasonable hope" of cure, but it reserved the label "terminal care period" for the "weeks or months immediately preceding death," when the patient "is so debilitated that he can no longer care for himself and/or is suffering from open lesions, drainage, or other conditions requiring nursing care." In the city of Chicago and the two Cook County suburbs included in the study area, that period lasted approximately three months.[5]

One goal of the study was to destigmatize advanced cancer. Over and over, the report insisted that the "disagreeable aspects" of the disease were extremely uncommon. Cancer patients with late stages of disease were no more likely to experience disfigurement or to exude unpleasant odors than were sufferers of diabetes, heart disease, and many other chronic afflictions. Slightly fewer than half of cancer patients in the terminal care period had even minor lesions and drainage, and good nursing care could eliminate any resulting odors.[6]

Researchers were able to determine the locations of 5,978 of the 6,236 cancer deaths in Chicago in 1949. Forty-three percent (representing 2,543 individuals) occurred at home and 56 percent (3,253 individuals) in general hospitals. The deaths of the remaining 182 individuals took place in other medical institutions.[7] The large number of hospital deaths may seem surprising in light of hospitals' efforts to restrict the number of beds occupied by people with fatal chronic diseases. "Dying is a process and death is an event," a British researcher reminds us. Although "there may be a tendency to confuse the venue of the care of the dying with the place where the death occurred," the "two are not necessarily coincident."[8] Many patients in the Chicago study had been forced to leave hospitals as soon as doctors declared they had little prospect of recovery. Although some hospitals also had policies prohibiting the entry of patients at the very end of life, those policies tended to be relatively ineffective. People who arrived at the emergency room in extremis often gained admission. The length of the final hospital stay tended to be relatively brief, a median of two weeks.[9]

Terminal Care for Cancer Patients found that more than a quarter (27 percent) of the patients dying at home did so because they could not find places in hospitals or other institutions. Nevertheless, the study concluded that lack of finances was a far more serious problem than the shortage of beds. A high proportion of terminal cancer patients had been compelled to relinquish paid employment prematurely. In addition, expensive medical treatment often had exhausted family resources.

Thirty percent of patients who died at home had needed institutional care but been unable to afford it.[10] Financial difficulties had forced other patients into substandard facilities. And many patients who remained at home could not pay for adequate assistance. Although physicians could not offer cures in late-stage cancer, they could help to alleviate pain and provide reassurance. Many hospitals, however, had too few social workers to ensure that patients received medical attention after discharge. Chicago's two government facilities were especially likely to send individuals with terminal cancer home without making plans for follow-up care. Investigators found "desperately ill" patients from both institutions who lacked access to physicians and therefore took taxis or private cars back to the hospitals to obtain essential dressings and medications.[11]

A sizeable proportion (43 percent) of terminal care patients required only the housekeeping and routine nursing services family and friends could provide. But some lived alone or with relatives who worked outside the home. And some needed skilled nursing care. Here it is helpful to recall Minnie Radamacher's 1914 letter from Long Beach, California, describing the services her niece Clara had provided her mother, Ida, in the final stages of cancer. Although the "terrible odor" from Ida's wound was almost unbearable, Clara cleaned it "3 and 4 times a day."[12] By the 1950s, however, most families assumed that only trained nurses had the expertise necessary to clean and dress wounds. Technological advances also had made other types of professional assistance essential. Nine percent of terminal cancer patients in Chicago had colostomies, indwelling catheters, or drainage tubes; and 1 percent had special feeding procedures.[13] Unfortunately, the nurses available to the poorer patients frequently lacked the qualifications necessary to provide those services.[14] And although the Visiting Nurse Association of Chicago furnished care to many patients who could not pay, the director acknowledged that advanced cancer patients often received fewer hours than were needed.[15] Both private and government funds to help low-income patients cover the cost of medications, dressings, and sick room supplies also were sparse.[16]

The Chicago report noted that because the characteristics of the study population were similar to those of terminal cancer patients elsewhere, the findings could be generalized to other parts of the United States and especially to other large metropolitan areas.[17] The records salvaged from New York's Presbyterian Hospital thus can help to humanize the Chicago numbers.

Located in the Washington Heights section of upper Manhattan, Presbyterian was founded in 1868. In 1911, it became one of the nation's first teaching hospitals by making arrangements to coordinate "the care of the sick with the educational and research program" of Columbia University's College of Physicians and Surgeons.[18] By the 1940s, Columbia-Presbyterian Medical Center included not just Presbyterian Hospital but also Babies Hospital, the Institute of Ophthalmology, the Neurological Institute, Sloane Hospital for Women, and Vanderbilt Clinic (an outpatient facility), all affiliated with the Columbia College of Physicians and Surgeons, New York Psychiatric Institute, and Columbia University School of Dental and Oral Surgery. As part of a large and prestigious academic medical center, Presbyterian drew patients from the Bronx, Riverdale, Long Island, and New Jersey. Most patients, however, came from the immediate neighborhood, long filled with Jewish, Irish, Greek, and Armenian immigrants and now attracting a growing number of African American and Afro-Caribbean residents.[19]

An enormous advantage of the Presbyterian Hospital records is that they included not only clinical notes but also extensive social workers' reports, often spanning months or years. Presbyterian first hired social workers in 1904 to teach student nurses "the social aspects of nursing." In 1921, the hospital established a separate social service department.[20] Although the workers repeatedly complained about understaffing throughout the following decades, they were able to serve a total of 17,527 patients in 1958.[21] Writing in the language of the day, the director defined the department's mission as providing "an essential link between care of the patient in the hospital and his restoration to an economic role in society."[22] Not all discharged patients, however, could return to productivity. Social workers thus devoted considerable time and energy to "the distracted family members who do not know which way to turn to make plans for the terminally ill patient."[23] The workers addressed several problems the Chicago study identified, and many appear to have done so with compassion and sensitivity. But visits with the social service department also meant that clients were subjected to its scrutiny. Distraught, impoverished patients and family members often violated the social workers' code of proper behavior.

Many patients had interacted with the social service department long before they needed terminal care. Because James Green, a 44-year-old African American was "very discouraged," social workers had seen him periodically throughout his hospital stay.[24] Workers had visited other patients to discuss their ability to pay hospital bills. Cancer frequently had created serious economic problems. Thirty-five-year-old Jonelle Harrington had earned a large salary as a fashion director in a major department store prior to her diagnosis with inoperable cancer of the vagina. By the time of her social work appointment, she had not worked for several months and planned to resign shortly. Her only relative was her 82-year-old father, a "leg amputee," who depended on her weekly financial contribution. Although Blue Shield would cover some of the bill for her lengthy hospitalization, she would have to go into debt to pay her share.[25]

The rapid spread of hospital insurance plans had resulted in a large increase in the proportion of patients in private and semiprivate rooms, but teaching hospitals continued to admit large numbers of indigent patients into the wards; and it was these patients who were the primary focus of the social service department.[26] Social workers often saw ward patients prior to admission to discuss their eligibility for "charity" beds as well as their ability to contribute to the cost of care. (Ward beds in many hospitals were entirely free, but Presbyterian charged a small amount to those who could pay.)[27]

Many ward patients were in severe financial straits. African Americans and Afro-Caribbeans reported especially desperate situations. Donald Turner, for example, a "tall and well built" man who was "not noticeably ill-looking," had pancreatic cancer. Since exhausting his annual thirteen days of sick leave from his post office job, he and his wife had "been living from hand to mouth relying on help from neighbors. They have no savings and bills are mounting up. They already owe two months' rent, a total of $64.40 and the landlady is beginning to clamor for payment. In addition, there are electricity and gas bills of $21 and a telephone bill of $16." Although Donald's wife previously had found employment as a domestic, "leg trouble" had prevented her from working during the past eight months.[28]

Thomas Norris, a middle-aged man with esophageal cancer, was appreciative of the social workers' "interest, especially around the subject of his wife and what is going to become of her during his illness. Neither he nor she had expected that he would be unable to work for any length of time." He had driven a truck in New Jersey railroad yards, earning $33 a week, but he now faced a long sickness, and the monthly rent of the apartment was $55 a month. After his death, his wife found work as a domestic and took in boarders.[29] Mabel Holmes, a 60-year-old woman, was "living on the edge of dependency" when she entered the hospital for a mastectomy. Her husband had been a shipping clerk, but he now was elderly and worked only two hours a day for half his salary ($7.50 a week). The couple paid $38 a month rent and had "no insurance whatever." After they paid Mabel's hospital bill, they would have "nothing left."[30]

In the great majority of cases, doctors pronounced patients terminal because further treatment would be futile. Two patients, however, received that label when they rejected recommended operations. One, Lily Freedman, was a 71-year-old woman who had been ill for months with cancer of the bile duct. At the advice of her physician, she had agreed to come to the hospital, but she "was loath to accept the idea of operation as it would 'only be a lot of misery' and would not make her any better. She was 'through' and waiting for the end of life."[31] The "terminal period" discussed in the Presbyterian Hospital records encompassed a much longer time frame than the one examined in the Chicago study. Many patients were still "up and about" and did not need skilled care. Four were able to return at least briefly to their jobs after discharge from the hospital.

The secrecy surrounding cancer complicated the task of arranging terminal care. "Pt. is not aware of his Dx, however his wife is aware of what is going on," read a physician's notes for a man with pancreatic cancer who was convinced that gall bladder trouble explained all his difficulties.[32] A social worker's report about another case was somewhat different: "The patient was not informed of the diagnosis. It seems likely, however, that she does know she has cancer, indeed she actually verbalized this fear to social service."[33] But even when patients realized the meaning of their symptoms, hospital staff tried to maintain the pretense that the problem was far less serious. And virtually no patients were told that their illnesses would end in death. Although Lily Freedman declared herself "through," most people who would not recover left the hospital expecting to get better.

Social workers thus typically made terminal care plans in consultation with family members alone. "Family members" occasionally was a capacious terms. The majority of people who met with social workers were spouses, siblings, or adult children, but a few had more tangential relationships to patients. Although Philip Andropov's sister-in-law was "interested and helpful in planning for the patient," she noted "that they had had slight contact during the past five years or so, and [that] his brother, to whom she was married has been dead for more than ten years."[34] Paul Castillo had no relatives in the United States; instead, the social worker consulted with a man who came from the same district in Spain and had remained a close friend.[35] In what follows, however, I refer to the people who made arrangements for terminal care as "relatives" or "family members," regardless of their relationships to the patient.

Most family members arrived at the social service department soon after learning that a relative's disease was fatal. Institutional demands rather than family and patient needs and desires had dictated the timing of disclosure. Up until that point, relatives had been able to hope for the best; in some cases, medical staff may have encouraged optimism. Now everything had changed. Because administrators insisted that terminally ill patients leave the hospital as soon as possible, medical staff or social workers could no longer conceal the horrifying prognosis from relatives. One social worker praised Andrea Becker, who took the news that her 47-year-old husband John had inoperable stomach cancer with metastases calmly, although "she had had no intimation before she saw the doctor . . . that the patient's condition was so serious." John's cantankerousness compounded her problems. In the hospital he had been "difficult to handle, refusing treatments and medication," and he insisted on going home. Andrea indicated that her husband had "always been stubborn and difficult to live with" and admitted that she did not have "an easy time ahead of her." Fortunately, however, she seemed like someone who could "take it."[36]

But not everyone responded so well according to the social workers' standards. William Pierson had entered the hospital "with an undiagnosed condition of the abdomen" that had "caused so much discomfort that he [had] been unable to work most of the time during the past six months." After performing an exploratory procedure, the doctor informed William's wife Margaret that the "growth" was inoperable. "Although her "manner was quite controlled, she expressed a good deal of bitterness over the fact that surgery had nothing to offer patient in spite of 'all the advertising of the societies for cancer research.' Margaret wondered what the use was of going to a doctor at the first sign of disease, when all the careful studies patient had in recent months showed nothing wrong with him at all."[37] The brother and sister of another patient were still "very upset" about the prognosis that the doctors had explained to them the night before. The social worker had to contact one of the doctors to talk with them again before they could respond "in a more realistic way."[38] A mother violated reigning gender norms while openly exhibiting anguish and despair. She and her husband had just learned that their 6-year-old boy had an inoperable brain tumor. The social worker found them "extremely anxious," noting that "their anxiety manifests itself in concern over a great many problems which are minor to patient's illness. The mother did all the talking and seemed to dominate the situation and when her husband played with his finger nails she reprimanded him, turning around and saying, 'stop that!' "[39] Another social worker accused a husband of being insufficiently "fatalistic," the opposite of the charge early-twentieth-century reformers leveled against immigrant mothers.[40]

Family members had to absorb not only a terrible prognosis but also the news that the hospital insisted on speedy discharge. Some could not understand how a facility they had trusted to help could abandon gravely ill patients. Although social workers explained that the hospital provided only acute care and had no place for people who could not benefit from further treatment, some family members may well have felt betrayed. One Presbyterian physician complained to the social service department that the husband of a woman with terminal disease had arrived at his office "demanding that she be kept here 'till well.' "[41] Relatives of patients who had spent long periods of time in the hospital found the notion of discharge especially difficult to accept. Pauline Borden, a 13-year-old girl with bone sarcoma, had arrived at Presbyterian to have her leg amputated. Because the wound became infected, recovery was slow, and Pauline remained at the hospital for two months. Such a terrible disease afflicting a junior high school girl had aroused the sympathy of the social service department, whose workers had carefully followed her progress. A few weeks after the surgery, they reported that "the child . . . is quite happy and not too uncomfortable. She gets a considerable amount of attention from the other patients and from the personnel" and that she "seems to have adjusted much better to the hospital than had been expected in her early days." Three days before her discharge, the workers noted that she "had settled down to the routine of hospital life and generally . . . been very happy and contented." Despite "moments in which she had been somewhat gloomy and depressed," she had generally "been very happy on the ward and amuse[d] herself by writing songs and parodies about the hospital routine and personnel." Six months later, when the parents brought Pauline back with a diagnosis of "generalized metastases," they assumed she would again be integrated into the ward; instead, they were instructed to make alternative plans for her care.[42]

Because ward patients represented the primary social work clientele, the Presbyterian Hospital records tell us little about private patients. It is likely, however, that people who could pay either out of pocket or with the help of insurance coverage were not under the same pressure to depart. Many had their own physicians, who could convince the hospital administration that longer stays were justified. Private patients also contributed to the economic well-being of the institution no matter how long they remained.

But terminally ill ward patients, too, occasionally were allowed to stay. Nine days after telling Pauline Borden's parents that she had to depart as soon as possible, the hospital reversed its decision. The social service record explained that her attending physician had "a chance of getting some special material from Cleveland," which would be "used on a research basis." Pauline thus remained in Presbyterian until her death, three months later.[43] Because Presbyterian was a major research hospital, her case probably was not unique.

Seven other patients died on the ward before plans could be put in place for their terminal care. One case was especially dramatic. Thomas Norris, the patient with advanced esophageal cancer, had seemed fine when his wife and friend arrived during visiting hours. He "was smiling and talking as he received them, saying he felt much better. After eating one bite of ice cream, which was brought in to him, he spat up a little blood. He then made a remark about 'Oh, Lord, help me! I'm dying!' He then coughed up a large quantity of blood and went into paroxysm and he became unconscious." The nurse "applied artificial respiration but could not get a pulse from the time of the hemorrhage, and he soon stopped breathing."[44]

The great majority of terminally ill ward patients did leave. In a few cases, social workers and relatives quickly agreed about the type of care required. Debilitated patients who were too poor to afford nursing services at home and lived alone or with people who worked during the day often had no option but to enter an institution. Placement was also clearly imperative when bedridden patients resided in rooming houses or walk-up apartments. More commonly, however, relatives assumed they would bring the patients home to die, even though social services strongly advocated institutionalization. In some cases, social workers' middle-class values may have led them to conclude that patients' relatives lacked the intelligence and sense of responsibility needed to provide care and that their homes were too "shabby," "untidy," and "congested." The four medical social workers who wrote the 1945 Boston study believed home care was appropriate only when the patient had "a room to himself—light, airy and equipped with a comfortable bed" as well as separate bathroom. "A peaceful and serene bedroom with reasonable quiet" also was considered essential.[45] Those criteria clearly excluded the apartments of the great majority of poor patients.

Of course, the Presbyterian hospital workers also were intimately acquainted with the serious problems encountered by all low-income families tending advanced cancer patients at home. And the workers knew all too well the common trajectory of the terminal stage of the disease. As the patient's condition deteriorated, home care would become increasingly difficult.

Nevertheless, it was far easier to recommend placement away from home than to find available beds. As in Boston and Chicago, few appropriate and affordable institutions existed. The Presbyterian Hospital social service department complained in its 1954 annual report that "planning for patients was seriously affected by the lack of suitable nursing homes, chronic care facilities, [and] terminal care facilities."[46] Although social workers provided families with lists of possible facilities and offered to write application letters, many places had no vacancies. Free beds were especially sparse. Social workers thus often sent applications to various institutions, explaining why the patients were particularly worthy of admission. Many requests went to homes for incurables, including the two facilities established by Rose Lathrop Hawthorne (St. Rose's Home for Incurable Patients and Rosary Hill Home) as well the House of Calvary, founded by an Episcopalian group in 1899. Social workers also frequently applied to two Bronx hospitals that had originated as homes for incurables, Montefiore and St. Barnabas Hospital for Chronic Diseases. Those institutions all had long waiting lists.

After the 1950 establishment of Francis Delafield Hospital, a city facility affiliated with Presbyterian and located nearby, many terminal patients who qualified as "indigent" according to Department of Welfare criteria found beds there. Throughout the 1940s, however, those patients were referred to the New York Cancer Institute, on Welfare (previously Blackwell's) Island. Despite its notoriously miserable conditions, that facility, too, was oversubscribed and could not accept patients on short notice. Patients who needed immediate care thus entered one of the municipal acute care hospitals instead. But even there, admission was not always automatic. One Presbyterian doctor urged a man in the last stages of pancreatic cancer to go to Harlem Hospital while awaiting a vacancy elsewhere. According to the social worker, "initially, both patient and his wife were against this but finally agreed." When the man arrived by ambulance, the hospital "refused to accept him on the grounds that they had no beds. They first told him he would need another operation and kept him waiting from 3 PM–11:30 PM before advising him to try Bellevue. Patient was finally admitted there at about 3 AM. His wife says that at that point he didn't care what happened to him."[47]

The social workers rarely applied to private acute care hospitals. When the daughter of a man with end-stage melanoma reported that a family doctor had suggested that the patient go to another private hospital if Presbyterian "kicked [him] out," the social worker "encouraged" the woman "not to become overly excited about that possibility" because most "private hospitals will not accept patients who need only nursing care."[48] Catholic hospitals were reputed to admit many patients others had rejected, and at the request of one wife, Presbyterian's social service department arranged a man's transfer to Misericordia Hospital, a small Catholic facility in the Bronx. But Misericordia imposed its own restrictions. Because the hospital "would not accept [the] patient for admission for ward care," the woman agreed to pay privately. "She felt that even if care was expensive the patient should have the best of care in the days that remain to him."[49]

The social service department urged several relatives to consider nursing homes, the new institutions that arose in the wake of the 1935 Social Security Act. Seeking to curtail the use of almshouses, that law stipulated that blind and needy people could receive federal aid only if they lived at home or in private institutions.[50] Between the early 1930s and 1950, the almshouse population plunged by nearly half, from 135,000 to 72,000.[51] Observers soon reported, however, that the problems of the almshouse had been transferred to nursing homes, which proliferated especially rapidly after the war; by 1953, there were 14,000 throughout the country.[52] Although most residents suffered from serious chronic conditions, medical attention was sparse and inadequate. Buildings were dilapidated and often unsanitary.[53] After the son of one Presbyterian Hospital patient reported in 1945 that he had visited several nursing homes "but did not find any of them satisfactory," the social worker commented, "It is true that the nursing homes with moderate prices leave much to be desired in the way of medical care and cheerful surroundings."[54]

Twenty years later the department was still complaining about "the lack of suitable and sufficient nursing homes." Any "patient who can afford and wishes more personal service and a really comfortable and attractive setting cannot be cared for in New York City." Although better facilities existed in New Jersey, Westchester, Connecticut, and Long Island, the "costs are high, particularly for a long period of time."[55] With few other options, however, social workers occasionally were forced to recommend nursing home placement, even in the city.

Some relatives who initially rejected institutionalization quickly reconsidered. Robert Nelson, the son of a man with an inoperable stomach cancer, lived "in Washington with his wife and two children in a 3-room apartment" and wanted to take his father to live with him. That plan "was eagerly picked up by the patient who is very fond of his son." The social worker tried to point out the impracticability of such an arrangement. Nevertheless, Robert was adamant. He "realized all the drawbacks but because of his strong feeling of wanting to be of assistance to his father who had sacrificed so much to bring him up and send him to college, the Washington plan was the one desired." A few weeks later, however, he sent a letter to the social worker explaining why it would be impossible to bring his father to Washington. The apartment was too small, his wife was busy caring for the two children, both of whom had asthma, his own health was poor, and the strain of witnessing his father's decline would be too great. As a result, he asked the social worker to help to make arrangements for the man in a chronic care facility outside New York.[56]

Sam Schneider's daughter Sadie also rapidly acceded to pressure. The Schneiders had lost everything in the Depression. Now Sam, his wife, Sadie, and a married daughter and her husband and two small children were "living in a four-room apartment of a four family house that Mr. Schneider formerly owned." Sadie first resisted the suggestion that her father go to an institution but then reconsidered, stating that she did "not see how the family can take care of him under their present living conditions" and did "not believe if he once was discharged home they would ever succeed in obtaining his consent to his transferral to a chronic hospital if his condition became worse." Sadie, too, asked the Presbyterian social worker for help in finding an institutional bed.[57]

Twenty-one of the forty-nine patients discussed in this chapter went directly to other institutions after discharge from Presbyterian—eleven to private facilities and ten to public ones. Another twenty patients returned home, at least for some time. Their relatives provided various reasons for rejecting institutional placement. Some insisted it was premature. The wife of a man with "sarcoma with multiple metastases" was "confident that he would with bed rest and care recover sufficiently to return to work" because he "had real pluck and perseverance." Although she "was well aware of what might be expected for the future in patient's condition," she "wanted to keep things as they were for as long as they could."[58] In another case, a son had visited both Rosary Hill and the House of Calvary as possibilities for his father, but "when he saw his father . . . up and around, he didn't have the heart to place him in a terminal care institution."[59]

Transferring a family member from the hospital to a terminal care facility also threatened to breach the secrecy surrounding the diagnosis and prognosis. The daughter of a man with metastatic bladder cancer explained that it would "be a difficult task to persuade her father to go to a chronic home, as he has been encouraged so much by the family that he will get well." She feared that "if he learns or suspects his condition he will commit suicide."[60] A brother initially agreed to send a man with metastatic disease to a home for incurables but then reconsidered, noting that "because of the name the patient would be made worse by a transfer there."[61] As in Chicago, the poor quality of available facilities was another source of concern. After visiting several nursing homes, one daughter reported that the ones "she had seen were either too bad or too expensive for her father."[62] Another family member refused to consider the Cancer Institute because it was associated with "public charity and inferior treatment."[63]

And several relatives emphasized the importance of caring for their own. This was true in the case of Arturo Rossi, a 71-year-old Italian-American discharged from the hospital with a diagnosis of metastatic neck cancer. He lived with his wife in the home of his married daughter in Rockville County. The other child was a son who appeared at the social service department, describing "a very close-knit family situation . . . There was no question in the son's mind but that not only would his father prefer to remain at home even at this point in his illness, but that his mother would not have it any other way." Although the son wanted information about available resources should his father's condition deteriorate further in the future, he "felt very strongly that his mother and sister would want to care for his father and would be able to do this despite the gravity of the medical situation."[64] (Of course, the female members of the family may have felt somewhat differently.) It also is possible that some relatives were motivated primarily by a desire to prevent patients from feeling abandoned. After arranging for one man's transfer to Delafield Hospital, a social worker commented, "He seems willing to accept any further hospitalization but I had the feeling that he felt that his family were rejecting him by not being able to arrange to take him home."[65]

The Presbyterian social service department furnished various types of assistance to terminally ill cancer patients who returned home after discharge. Social workers told one woman that she could find gauze for her husband's wound care at Macy's, supplied blankets and sheets to enable a man to sleep on the couch so his ailing wife could be more comfortable in their bed, and gave many other relatives advice about transportation back to the hospital for clinic appointments. Because Donald Turner doubted he would be able to return to his post office job, which involved "a lot of physical effort in lifting heavy mail bags," his social worker phoned his supervisor to request that Donald be given "suitable light employment." When advancing disease forced Donald to relinquish his job and apply for welfare, the worker urged that his allowance be increased "to cover telephone costs in case of emergencies."[66] Like many family members described in the Chicago report, Ellen Cramer's 20-year-old daughter "had never cared for a sick person and so knows little about it." Her social worker thus made arrangements with the Visiting Nurse Service to provide assistance.[67] Other social workers contacted various registries of private duty nurses to find ones available to care for patients who did not qualify for the visiting service.

But the social workers were aware that the services they could furnish were grossly inadequate, and as patients' health declined, the workers continued to press for institutional placement. It was possible to determine the place of death for fourteen of the twenty patients who returned home from Presbyterian. Seven died at home and another seven in institutions.

One of those who died in an institution after a long period at home was Mabel Holmes. Mabel first entered Presbyterian Hospital in February 1942, "accompanied by her husband who seems attentive to her and anxious that she should have proper care." The social worker had been immediately impressed by their 40-year marriage. When the husband, Louis, learned that Mabel would have to undergo a mastectomy and that "only time would tell whether the trouble has been eradicated," he remarked, "She is all I have" and stressed "the happiness they have had together."[68]

Mabel recovered well from her surgery. Although she was initially "quite timid about going home," Louis was "devoted to her" as well as a "good cook & very handy about the apt." Visiting the home a month after Mabel's discharge, the social worker found her "considerably stronger." At a clinic appointment in June, she seemed happy and had "regained her pep & poise." The following month, however, she began "to show some signs of slipping," and the social worker told Louis that the cancer had returned. By December, Mabel was "slowly losing ground," though "still able to be up & fairly active." In this case, the social worker initially hesitated to press for institutional care because "it would be a grief to these two old people at this time. They have had a very happy long married life." But then, in March 1943, Louis was hospitalized for ten days following a heart attack, and the social worker decided institutional placement was essential. Should Mabel become bedridden, it was "questionable whether her husband in his present precarious state of health could be trusted with her care."

In July, Mabel had "a surprising amount of energy" and was still active. In September, however, Louis wrote that she had been found wandering on the street, her mind "a complete blank." A few weeks later she had a second "attack." On November 10 the social worker visiting the home saw Mabel "sitting up in bed and somewhat breathless." She also was incontinent. Louis had to stay with her constantly and was assisted by a woman who came in daily to bathe her. Although he still insisted she stay home, the social worker sent an application to the House of Calvary, in case he changed his mind. Three days later, he agreed.

Now the social worker herself expressed qualms, warning Louis "that patient may not be satisfied and that visiting hours are only twice a week." He responded that caring for her at home was "becoming too difficult," and that it was "increasingly difficult to get help." Mabel left by ambulance on November 16. Later that day Louis wrote a letter thanking the department for all its help. "Badly as I hated to see Mabel taken from her home, events took shape that I realized such was the best and humane step for both of us. She had developed an unconscious mind and had I attempted to keep her home longer the result may have been more serious. Although I was perfectly willing to make any sacrifice for her good and comfort there are certain cares needed for a bedded woman which only another woman can do." The sisters at the House of Calvary had "received her very nicely," and he was "sure they will do everything possible to make her comfortable." She died there on November 19, three days after her arrival.

Returning to Presbyterian rarely was an option. In some cases, doctors had inserted notes in the case files stating that the patients required only nursing care and therefore should not be allowed to reenter. Those who tried to gain readmission often failed to do so. Carla Alessi, a 42-year-old Italian woman with five children, had entered Presbyterian in February 1943, complaining of fever and bladder problems, and soon received a diagnosis of carcinoma of the cervix. Although the disease was too advanced to be removed by surgery, she remained eighteen days to undergo radiotherapy treatments. She refused the operation the doctors recommended to relieve the pain, however. In mid-August, she again applied to the hospital, "complaining bitterly" of pain in her left back and thigh and asking for the operation she previously had rejected. The social worker informed the doctor "that although there were some negative factors in the rather poor family relationships, low income, hostility of husband, etc., a plan could probably be worked out with family so that she would be ready to leave [the hospital] when doctors were prepared to discharge her." Nevertheless, the doctor decided the risk was too great that Carla would have no place to go after the surgery and that the hospital would be forced to "tie up one of the beds with a chronic patient." As a result, he refused to grant her permission to return for the operation.[69]

A few months later, Presbyterian denied a family doctor's request to readmit Mabel Holmes. A year and a half after she returned home, the doctor phoned her social worker at the hospital, "explaining that the patient is now quite sick, is vomiting and feels weak." The doctor said that she needed hospital care because the cancer was "now pretty much generalized." After conferring with the woman's physician at Presbyterian, the social worker left a message for the family doctor stating that Mabel could not be "admitted to our ward at the present time and that an application should be forwarded for the patient's admission to the House of Calvary." If Mabel's case was "emergent," she should go to the local city hospital.[70]

Although cancer patients who arrived at Chicago hospitals close to death often gained admission, Presbyterian's emergency room rejected some dying patients who tried to return without asking permission. One such patient was Paul Symanski, a 53-year-old Polish man who had been receiving radiotherapy for metastatic tongue cancer at the hospital out-patient clinic. Trying to plan terminal care for him, the social worker had attempted to contact his daughter, with whom he lived, but the daughter worked at Macy's and could not be reached during the day. The only option, therefore, was to talk to Paul directly. But he "would not accept the idea of terminal care." He had "washed cars for a living for the past 22 years and still [did] this at night (6:00 PM to 3:00 AM)." He had no intention of quitting because he felt that the job "gets his mind off his troubles as well as giving him an opportunity to earn his living and do something he enjoys." Therefore, no plans were in place the night Paul took a sharp turn for the worse and arrived at the emergency room by ambulance. The physician on duty angrily noted that because the patient had "only terminal care as a future," the social service department should have made suitable arrangements before then. Paul went by taxi to Francis Delafield Hospital, where he died ten days later.[71]

Although the records do not indicate how these three patients responded to their rejection by Presbyterian, the Boston study noted that "many a [terminal cancer] patient" who initially returns home "seeks admission to the hospital where he has previously received treatment," as his condition worsens. "A sick person finds it difficult to accept any interpretation regarding hospital policy or differentiation of general and chronic hospitals. 'Doctor, why are you telling me to go to another hospital [or to a nursing home] when I am sick and need care?' is a frequent question, and a patient groping and searching for relief of pain feels only that the hospital is withdrawing help."[72]

One Presbyterian case ended differently. Two weeks after a boy with a brain tumor was discharged from the hospital, the mother phoned the social service department, stating that her son had had "continual headaches since his discharge but this morning had severe pain in his head, became rigid for about a period of 5 minutes, and vomited violently, after which he seemed stuperous." The mother asked the social worker to contact the doctor to ask if she should bring the boy to the emergency room. The doctor replied that he had informed the parents of both the diagnosis and prognosis and that "they understood that the patient could not live and might die at any time." The social worker then "called the mother and told her that the doctors were not surprised to hear of the patient's condition, that these attacks would probably occur again but that no medication was necessary, pointing out that had the boy been in the hospital nothing more could be done than she was doing in her sympathetic, loving way." When the mother replied that she understood the situation but felt "helpless," the social worker suggested she phone her family physician. Soon afterwards, the boy returned to Presbyterian for a regularly scheduled procedure. When that was finished, he again went home. A few days later, he arrived at the emergency room "in extremis." It is possible that the mother agreed to pay privately. Or perhaps the hospital reversed its previous decision because the boy had so recently left the hospital. In any case, this time he was admitted. He died the next day.[73]

If the Chicago and Boston reports revealed the disastrous consequences of sending very poor, terminally ill cancer patients home without adequate supervision, the Presbyterian Hospital case files demonstrate the benefits of having access to a social service department. Presbyterian social workers referred families to suitable terminal care facilities and often wrote application letters explaining why the patients merited admission. When patients returned home, the social workers provided caregivers with crucial information and supplies. But no social worker could compensate for the paucity of services for this vulnerable population. In New York, as in Chicago, many terminal cancer patients without adequate financial resources ended their days in substandard nursing homes and public chronic care facilities. Others received care from overburdened family members bereft of essential assistance.

The two city reports and the Presbyterian Hospital patient records also enable us to revisit Philippe Ariès's contention that relatives welcomed institutionalization as a way to distance themselves from the dying process.[74] Although one of Robert

Nelson's objections to providing a home for his father was the fear of witnessing his final deterioration, most families regarded institutional placement as a last resort. Several continued to care for family members at home for weeks or months after social workers had suggested institutionalization. A major reason several patients died within days of entering terminal facilities was that their families had waited so long to surrender them. But what made relatives finally conclude that institutional placement was imperative? According to the Chicago report, in many cases "the family can care for the patient at home during the earlier part of his terminal care period, and do so gladly, but find themselves unable to continue as the end approaches, the patient requires more intensive care, and their own strength is being exhausted."[75] In an application letter to the House of Calvary on behalf of one former Presbyterian patient, the social worker emphasized that his pain had increased and that five months of caregiving had overwhelmed his wife, whose own health had declined. There was an additional factor in Mabel Holmes's placement in that facility: her husband considered it inappropriate for him to deal with her incontinence. (It is possible that a wife would have made a different decision.) Although the social workers did not mention family members' distaste for other signs of end-stage cancer, that, too, often may have been important. The Chicago report noted "a greater tendency toward institutional care" of patients who presented "more unpleasant manifestations of the disease."[76]

And yet, families did not simply banish dying people to institutions. Most wanted to maintain the contact that would show patients they remained special and valuable even as the end drew near. The Boston report noted that 125 patients (63 percent of the total sample) stated that they would have accepted placement in a chronic hospital if it had been located near home. But "during the period of study the only resources of this type available for immediate admission were chronic hospitals so distant from Boston that the families could visit them only occasionally, with much cost in time and money. Of the 212 patients who died in these hospitals, 13 were without close ties with family or friends, and therefore felt no objection to entering a hospital at some distance from their place of residence. It was believed that the other 8 entered only because of necessity for care and contrary to their own choice."[77]

Access also was a critical concern for relatives of Presbyterian Hospital patients. One family, for example, rejected Rosary Hill, in the center of Westchester County, in favor of the House of Calvary, close to their home in the Bronx. There, family members would be able to "visit the patient frequently."[78] Others used scarce resources to pay for the private beds that allowed more liberal visiting privileges. Applying to a terminal care facility on behalf of a man with metastatic lung cancer, the Presbyterian Hospital social worker noted that the daughter-in-law "expressed the wish that the patient could be treated [in a] semi-private room which would enable them to visit him for frequently; she felt that for the short remaining period of his life they could manage this expense."[79] But that cost was prohibitive to many families. We can assume that Louis Holmes would have liked to remain by his beloved wife after she went to spend her last days in a House of Calvary free bed. The social worker sadly explained, however, that he would be able to see Mabel only for short periods twice a week.[80] In most instances, enrolling relatives in terminal care facilities meant sending them away to die alone.

CHAPTER 8

The Sacred and the Spiritual

By the end of World War II, attitudes toward dying had changed considerably from the sentiments expressed by Mary Ann Webber when she wrote from her Vermont farmhouse to remind her children that "the seeds of death are within us" and urge them to "be prepared to meet the event with Christian fortitude." We saw that tuberculosis occasionally was idealized in the nineteenth century because it provided time for spiritual reflection; now a good death was widely believed to be quick and sudden.[1] As a secular model of *ars moriendi* increasingly replaced the older, religious version, acceptance of God's will had become synonymous with resignation and fatalism. Popular culture encouraged patients to pursue every available medical therapy in the fight for recovery and ignore the signs of encroaching disease insofar as possible. Many people confronting terminal disease followed that advice.

Nevertheless, scientific optimism had not extinguished the spiritual component of death and dying. Even during the 1950s, at the height of popular confidence in medical prowess, most Americans belonged to religious congregations and believed in God and an afterlife. And many who eschewed religious institutions and beliefs looked to other forms of spirituality to try to understand the tragedies medicine could not avert.

Because the United States had become a far more religiously pluralistic society during the postwar period than it had been a century earlier, patients and caregivers drew on extremely diverse faith traditions. By the late fifties and early sixties, those included many eastern religions, such as Buddhism, Hinduism, Sikhism, Taoism, and Hare Krishna.[2] Long before her daughter Amanda fell ill, Dorothy Dushkin had begun to study the writings of Krishnamurthi, a spiritual leader from India, and Hubert Benoit, a Frenchman who helped to introduce Buddhism to the West. As she struggled to come to terms with her daughter's grim outlook, she read and reread the works of both writers, redoubling her efforts to incorporate their insights into her life. She even addressed a query to Benoit via his American publisher, and when Benoit sent a reply "that was not at all perfunctory," she was overjoyed, and they began a brief correspondence.[3] She also tried to read Alan Watts, an American Buddhist writer, before concluding he was "too wordy."[4]

Eleven months before Amanda died, Dorothy stressed the importance of remaining focused on the present. "So much of this grief is concerned with time," she wrote in February 1961. "Waiting for developments, for diagnosis, for cure or defeat, endless speculation as to becoming—imagination feeds on time." The only solution was "drastic & fundamental change—a life in the present so complete that past & future are unimportant."[5] The injunction to pay strict attention to painful emotions also proved useful. "The only peace of mind I get these days of watching Amanda's desperate illness," Dorothy wrote on December 12, 1961, "is by plunging into the confusion of my reactions." And a month later she was able to find at least a modicum of serenity while awaiting Amanda's death: "I feel no bitterness or need for inveighing against fate. This is only our human condition—common to all. I can only yield to this suffering without need to blame, resent or be defeated. It has come and that is—just as clouds come over the mountain."[6]

John Gunther's ex-wife Frances also adhered to a faith tradition that differed significantly from the Protestantism that had dominated the United States throughout the nineteenth century. John had resisted including "A Word from Frances" as a postscript to *Death Be Not Proud* and did so only at his publisher's insistence. Although he acknowledged the section was "beautifully, movingly, exquisitely written," he also pronounced it too personal. "I tell a story," he remarked, "and she tells of a relationship and how you stand it when a relationship is broken by something external."[7]

Frances described in lavish detail the myriad activities she and her son had enjoyed together and poured out her anguish. "All the things he loved tear at my heart because he is no longer here on earth to enjoy them."[8] Guilt also assailed her. "Missing him now, I am haunted by my own shortcomings, how often I failed him," she confessed. "I think every parent must have a sense of failure, even of sin, merely in remaining alive after the death of a child. One feels that it is not right to live when one's child has died, that one should somehow have found the way to give one's life to save his life. Failing there, one's failures during his too brief life seem all the harder to bear and forgive."[9]

But perhaps John objected above all to the deep sense of the sacred and spiritual Frances expressed. Frances's background was Judaism, a religion that concentrates far more on living a good life than on preparing for any kind of eternal existence.[10] In a "Word from Frances," she criticized the "single theme" dominating condolence letters from friends and acquaintances—"sympathy with us in facing a mysterious stroke of God's will that seemed inexplicable, unjustifiable and yet, being God's will, must also be part of some great plan beyond our mortal ken, perhaps sparing him or us greater pain or loss." Although she had always "had a spontaneous, instinctive sense of the reality of God" and had prayed continually throughout her son's long illness, she had not felt that "God has personally singled out either him or us for any special act, either of animosity or generosity."[11] She mentioned the possibility of immortality, but only briefly and vaguely, writing that death was "a part of Life, like Birth. Not the final part. . . . Only the final scene in a single act of a play that goes on forever."[12] Rather than imagining an existence after death, she focused on the present, exhorting her readers to live as intensely as possible and savor their blessings: "Today when I see parents impatient or tired or bored with their children, I wish I could say to them, But they are alive, think of the wonder of that! They may be a care and a burden, but think, they are alive! You can touch them—what a miracle!" The experience of terrible loss must lead to greater compassion, "obliterating . . . the ideas of evil and hate and the enemy, and transmuting them, with the alchemy of suffering, into ideas of clarity and charity."[13]

Both Gunthers received thousands of letters from readers. Those addressed solely to Frances came overwhelmingly from women and focused almost exclusively on her section. Many referred to God. "God bless you and grant that your suffering, and Johnny's, may help to enlighten this terrible world and ease some of its pain," wrote Lillian R. Lieber of Brooklyn. Dora S. Raue, from Chestnut Hill, Massachusetts, assured Frances that "the Ruler of the Universe is a God of Life, not of Death. Nothing dies." Irene Gromik, a Detroit nurse, wrote: "Your belief in God and in prayer undoubtedly gave him the world's greatest gift, a peace of mind." A woman who identified herself only as a mother told Frances that she was "one of God's chosen people."[14] Others sent prayers or suggested religious writings she might find helpful.

Letters which failed to mention God often praised Frances's spiritual insights. "Your spiritual strength has found grace of phrase and expression," wrote Mrs. Willard VanHazel of Chicago. Sophie A. Udin wrote from New York, "The spiritual values which you have put into Johnny's life, tell of your spiritual depth and make me certain that your life will give much of this spirituality to others." Mrs. C. E. Morrison, in Canton, Ohio, wrote, "I especially admire your spirit—no bitterness in spite of your untold grief and suffering."[15] Many others promised Frances that her words would make them better people.

The most poignant letters came from mothers whose children either had died or currently faced life-threatening illnesses. At a time when most parents could expect their offspring to outlive them, those correspondents had felt profoundly alone in their grief. Frances's "Word" in *Death Be Not Proud* punctured their isolation. "Even since we lost our daughter Eleanor last August," wrote Lora J. Jackson of Madison, Wisconsin, "there seemed to be no friend who understood except, perhaps, Mrs. Frank Lloyd Wright who lost a daughter two years ago, and, now, you." Like Frances, correspondents complained that most words of comfort felt empty. "How I loathe the people who insist, 'You are a better, stronger person,' " wrote Georgiana A. Dillon, a Los Angeles woman whose 5-year-old son suffered from polio. "I don't want to be a stronger person," she continued, "I want an active, joyous, busy little guy on a bicycle. But I guess I'll have to settle for strength of character." Marianne R. Peters, a Chicago woman who had just learned that her daughter had diabetes, wrote: "It took your words to make me see my blessings. Yes, they have been pointed out by many persons but heretofore seemed to be only rationalizations. Now I thank God for the gift of just her life."[16]

Although Frances wrote little about life after death, her few thoughts about immortality resonated with several bereaved mothers. "Our Joan died of leukemia at the age of six, on October tenth, on a bright lovely autumn day after only being ill since May," wrote Eileen Bowen in Victoria, British Columbia. "As I stood in the hospital room . . . the thought flashed through my mind that I had never been so close to death and in the same instant was connected by something within, 'You have never been so close to life,' and through my mind coursed those same thoughts that you spoke of so beautifully and clearly, about death not being the final part—just another step." Elizabeth Custer Nearing, of Lakewood, Ohio, read *Death Be Not Proud* soon after the death of her son, another John. She was convinced that the spirits of both boys "are alive, more so than when they were encumbered with bodies."[17]

Sociological studies of hospitals reported that dying people, too, frequently turned to faith traditions for support. According to Barney G. Glaser and Anselm L. Strauss, some patients who learned that their illness was terminal began "immediately to prepare themselves for death through religion." Others found it "an easy transition to slip from philosophical to religious terms, a transition often aided by the chaplain."[18] Renée C. Fox's classic 1959 study, *Experiment Perilous*, is especially revealing. The ward she observed consisted of critically ill men who had metabolic diseases that were poorly understood at the time and who had agreed to participate in medical experiments. Although the frequent occurrence of death demonstrated that medicine was not as infallible as the men had hoped, they identified closely with their physicians as well as the research. Perhaps as a result, the men carefully concealed all evidence of religious observance. Yet "in a number of indirect and subterranean ways," the ward was "a deeply religious community."

Patients read their Bibles and prayer books and prayed the rosary, though only "behind the drawn curtains." Similarly, they listened to religious programs on "muted" radios. Because the religious component of the ward was so private, it took Fox a long time to perceive it, to "catch a glimpse of a cherished crucifix, for example, or hear a snatch of a radio-borne sermon." And only gradually did she realize "that the special treatment given to patients who were clergymen, the highly intellectual theological discussions among Protestant, Catholic, and Jewish patients, and some of the joking in which patients engaged were all oblique forms of worship."[19]

Other researchers appear to have ignored any evidence of persistent religiosity they uncovered. In his study of death in Schenectady, New York, historian Robert V. Wells analyzed every mention of mortality in the extensive correspondence between Lewis B. Sebring, a newspaper reporter, and his parents between 1923 and 1949. Because the family rarely used spiritual language to discuss the loss of friends, relatives, acquaintances, and celebrities, Wells argued that "in spite of being devoted church members, even the elder Sebrings abandoned all but the most fleeting reliance on religion when confronting death."[20] Wells's quotations from Lewis Sebring's letters describing the death of his 86-year-old mother, Agnes, in 1953, however, point to a different conclusion. Agnes entered the hospital expecting to recover but gradually realized the end was near. She used a visit from her pastor to acknowledge that "she did not think she would remain alive much longer." Although popular culture no longer demanded that relatives hear the last words of patients and discern their spiritual state, Sebring noted that during the hours she slipped into unconsciousness, Agnes "spoke often of church communion, especially the words of Christ: 'This is my body which is broken for you.'" And he drew comfort from his last glimpse of his mother. The evening before she died, when she was in a coma, he saw "a strange thing . . . one of those things you always remember. As she lay there, the rays of the setting sun shone upon her face, and seemed to light up her countenance in a holy glow—almost as though it were the last tribute of her Maker to a woman who had lived a fine life."[21]

For the remainder of this chapter, we will examine two personal narratives that describe very different spiritual approaches to catastrophic illness and impending death. Flannery O'Connor drew on the Catholicism she had learned as a child, while Rachel Carson relied on a form of spirituality she developed outside organized religion. O'Connor received a diagnosis of systemic lupus erythematosus at the age of 26, in 1951, and died of that disease in August 1964. Although Carson was considerably older when she was diagnosed with breast cancer, her illness overlapped with that of O'Connor. Carson was 52 in 1960, when she underwent a mastectomy. She died in April 1964, four months before O'Connor. As unmarried, highly accomplished, and extremely prominent women, neither can be considered a typical patient. Both, however, help us understand how patients could come to terms with mortality during a period of extraordinary faith in medical science.

Partly because O'Connor and Carson left voluminous correspondence and partly because their premature deaths in the midst of flourishing careers have special poignancy, their final illnesses have received considerable historical attention.[22] In addition, literary scholars have sought to understand how sickness shaped O'Connor's writings.[23] Although I have drawn on previous accounts, I have also departed from them in significant ways, to emphasize the ways the culture of the time influenced O'Connor's response to serious illness and impending death and to challenge the dominant portrayal of her mother's care. For Carson, I follow historian Robert A. Aronowitz's study in noting that she devoted enormous effort to concealing her deteriorating health from the public as well as from most friends and acquaintances. But I focus on her struggle to overcome her customary reticence and honestly communicate the experience of her failing body to the person who mattered most to her. Physician Rita Charon argues that the "telling of pain and suffering" enables patients "to give voice to what they endure and to frame the illness so as to escape domination by it."[24] In letters to an intimate friend, Carson both bore witness to her illness and transcended it.

Although life circumstances separated the two women, they had much in common. They received treatments that represented some of the most dazzling achievements of the new medical science but remained profoundly ambivalent about that science. Like many other contemporary patients, both also initially had difficulty learning the truth about their conditions. Intensely private women, O'Connor and Carson often colluded in that secrecy. Despite the culture of fear, evasion, and hostility enveloping death and dying in the middle of the twentieth century, however, both women found ways to approach the end of life with understanding and integrity.

Flannery O'Connor

Flannery O'Connor was born on March 25, 1925, in Savannah, Georgia, the only child of Regina Clark O'Connor, a descendant of a prominent southern Catholic family, and Edward Francis O'Connor, a businessman and lifelong Catholic. By the time Flannery was 13, her father had begun to display early symptoms of systemic lupus erythematosus, a painful, degenerative autoimmune disease. Flannery did not know that her father had lupus, a childhood friend later recalled. "She never had any idea. In those days, they didn't tell children those things."[25] Nevertheless, she could hardly have been unaware of the ravages the disease inflicted on her father's life. In 1940 he was forced to resign his job and retire to his wife's family home in Milledgeville, a small town in the center of the state. As biographer Brad Gooch writes, Edward's "fifteen-year-old daughter watched as the father she adored—a middle-aged man, otherwise in his prime—suffered a mysterious, painful, wasting death from the fatal illness."[26] He died in February 1941, shortly after his 45th birthday. One of Flannery's rare comments about that event appeared in her diary two years later: "The reality of death has come upon us and a consciousness of the power of God has broken our complacency, like a bullet in the side. A sense of the dramatic, of the tragic, of the infinite, has descended upon us, filling us with grief, but even above grief, wonder. Our plans were so beautifully laid out, ready to be carried to action, but with a magnificent certainty God laid them aside and said, 'You have forgotten—mine?' "[27]

By the fall of 1950, Flannery may have intuited that the same disease might derail her own plans. She had left Georgia first to attend the University of Iowa Writers' Workshop and then to reside at Yaddo, the celebrated artist's colony in Saratoga Springs, New York. Now she lived as a boarder at the Connecticut home of the poet Robert Fitzgerald and his wife, Sally, whom she had met at Yaddo. Although she tried to hide the severity of the pain and heaviness she felt in her arms and shoulders from the Fitzgeralds, she became seriously ill on the train back to Milledgeville for Christmas. The uncle who met her at the station saw "a shriveled old woman."[28] Seven years later, she fictionalized her homecoming even more dramatically in "The Enduring Chill." When the main character, 25-year-old Asbury, steps off the train, "the smile vanished so suddenly" from his mother's face, "the shocked look that replaced it was so complete, that he realized for the first time that he must look as ill as he was." She had glimpsed "death in his face at once."[29]

A physician at the local hospital diagnosed Flannery with acute rheumatoid arthritis and prescribed ACTH (adrenocorticotropic hormone). ACTH, a derivative of cortisone, was one of the major postwar medical successes. News of its discovery in a Mayo Clinic laboratory in 1949 inspired the hope that chronic diseases, like many acute ones, soon would succumb to new pharmaceutical treatments. The following year the principal investigators won the Nobel Prize. *Life* magazine entitled an article about it "The Miracle of Cortisone."[30] "Even though ACTH and cortisone may not fulfill all of the therapeutic expectations to which recent enthusiasm has given rise," declared George W. Thorne of Peter Bent Brigham Hospital and Harvard Medical School, "it is evident that they have opened a new era in medicine."[31] Flannery, however, was soon aware of what the public would only gradually learn. The drug produced serious side effects, including the rounding of the face and mental overstimulation. "One gets stirred by cortisone," wrote the Harvard professor and psychiatrist Robert Coles, who was an intern when hospitals began to use the medication. "The drug that was saving [O'Connor's] life was also, to some extent, stirring her body and mind."[32] The substance was also expensive. For many years, Flannery would use a significant portion of her earnings from writings and talks to pay for the medication.

In February 1951, Regina O'Connor transferred her daughter to Emory University Hospital at Atlanta, where the doctor realized that the troubles stemmed from lupus, another disease believed treatable by ACTH. Flannery now received massive injections of the drug. In keeping with the custom of the time, the physician disclosed the diagnosis to Regina alone. Soon after receiving the news, she phoned the Fitzgeralds to report that Flannery was dying of lupus but did not know she had that disease.[33]

When Flannery finally left the hospital in March, she was forced to adopt an invalid regime, giving herself daily shots of ACTH and adhering to a salt-free diet. Because she was too weak to climb the stairs of the Milledgeville house, she and her mother moved to Andalusia, a farm owned by Regina's brother a few miles away. Throughout the following summer, new exacerbations repeatedly sent Flannery back to the hospital. By June 1952, however, she felt well enough to return to her friends' Connecticut home. Soon after Flannery arrived, Sally Fitzgerald informed her that the disease was lupus, not arthritis. "That was devastating knowledge," Sally later recalled, "that she was going to have to live with uncertainty, that she would not be autonomous and independent. I didn't minimize what she would have to go through. . . . But I never really regretted it. I knew it was what Flannery wanted. The atmosphere cleared."[34] Although protection of individuals' emotional states was a key justification for concealing fearsome diagnoses, many wanted the truth. "I know now that it is lupus and am very glad to know," Flannery wrote to Robert Fitzgerald, then in Indiana.[35] In the future she always would insist on getting medical news "straight."[36]

A lupus diagnosis must have been especially terrifying to Edward O'Connor's daughter. Writing to Robert Lowell and his wife, Elizabeth Hardwick, Flannery insisted the affliction was now curable. "I have a disease called lupus. . . . My father had it some twelve or fifteen years ago but at that time there was nothing for it but the undertaker; now it can be controlled with ACTH."[37] But did she really believe she could escape her father's fate? As we have seen, Regina assumed her daughter would die despite the advent of a remarkable new medication. Flannery may have had her own doubts about medicine's ability to heal. In "The Life You Save May Be Your Own," one character asserts: "There's one of these doctors in Atlanta that's taken a knife and cut [out] the human heart . . . and held it in his hand . . . and studied it like it was a day-old chicken, and lady . . . he don't know more about it than you or me."[38] Many other stories discuss people facing death. In personal letters Flannery referred to her primary doctor as "the scientist," an honorific that her friends undoubtedly perceived as tinged with mockery. Although she often expressed gratitude for his help, urged other sufferers to consult him, and cautioned friends against self-medication, she shared little of her era's reverence for science.

And the medication could only partially control the disease. Although Flannery had expected to remain with the Fitzgeralds indefinitely, a viral infection soon compelled her return to Georgia. There, the doctor concluded that the virus had reactivated the lupus; to control the flare-up, he ordered blood transfusions and raised the ACTH dosage. Finally realizing that she would have to relinquish her dream of residing in the North, Flannery asked the Fitzgeralds to send back her belongings.

Because Flannery spent the remainder of her life in her mother's house and under her care, biographers and critics devote considerable attention to the mother/daughter relationship. Joyce Carol Oates summarizes the consensus when she refers to the "lifelong tug-of-war" that "seems to have been enacted between the (quietly, slyly) rebellious Flannery and the stubborn, self-righteous, and unflagging Regina, whose efforts to mold her daughter into 'the perfect Southern little girl' were doomed to failure."[39] But scholars base that conclusion largely on the observations of Flannery's friends who visited her at Andalusia.[40] They, in turn, may have viewed Regina through the veil of cultural anxieties about domineering and intrusive mothers. Philip Wylie warned of the dangers of "momism" in his 1942 *Generation of Vipers*. By 1955, when the book was in its twentieth printing, the American Library Association designated it one of the major nonfiction works of the first half of the twentieth century.[41] Psychological literature of the era also encouraged hostility toward strong mothers. Women who reared children alone, whether because they were widows (like Regina) or had never married, were especially suspect. Many experts blamed exorbitant mother-love for the high prevalence of mental illness among World War II servicemen. After the war, psychiatrists argued that veterans could readjust to civilian life only by separating themselves from their mothers. Above all, they must not return home.[42]

Five of Flannery's stories—"Greenleaf," "The Comforts of Home," "Everything That Rises Must Converge," "The Enduring Chill," and "Good Country People"—describe educated, urbane, adult children who live with widowed mothers. Perhaps seeking to underline the relationship between those fictional situations and the contemporary psychological discourse, O'Connor made the children in four of those stories male. Although the children relentlessly disparage the attitudes of their less intellectual mothers, the offspring are the primary objects of fun. Some have health problems that partially explain their reliance on their mothers; all are too passive to leave. Thomas in "The Comforts of Home" decides to depart because his mother is "about to wreck the peace of the house" by inviting a young woman to live with them. But Thomas is incapable of action. "He did not know where a suitcase was, he disliked to pack, he needed his books, his typewriter was not portable, he was used to an electric blanket, he could not bear to eat in restaurants."[43] The mothers support their offspring and place few restrictions on their lives, but the children are uniformly surly and resentful. In such stories, Flannery appears to ridicule and repudiate her condescending attitude toward Regina and express gratitude for her care.[44]

Regina's story deserves its own account. What was it like for her to welcome home a gravely ill 25-year-old daughter and care for her for the next fourteen years? Did she resent the restrictions and burdens Flannery's sickness and dependence placed on her own life? Convinced (at least at the outset) that medical advances could not reverse lupus's fatal course, how could Regina bear to witness her daughter's struggle with the disease? To what extent did she share Flannery's search for acceptance?

If Flannery's education and literary achievements separated mother and daughter, religion bound them together. Flannery's rebelliousness never included the Catholicism of her upbringing. Despite the rigid writing schedule she adopted at Andalusia, she found time for daily religious observation. Upon waking, she read prayers from the breviary by her bed. After coffee, she and her mother attended morning mass. Before going to sleep, Flannery often recited the prayers for compline or read from the works of Thomas Aquinas.[45] "I write the way I do because and only because I am a Catholic," she informed a Notre Dame teacher. "I feel that if I were not a Catholic, I would have no reason to write, no reason to see, no reason ever to feel horrified or even to enjoy anything. I am a born Catholic, went to Catholic schools in my early years, and have never left or wanted to leave the Church."[46] As we will see, Flannery also drew on her faith as she struggled to come to terms with her deteriorating body and approaching mortality.

New problems continued to arise. "I have a decided limp," she confessed to a friend in February 1954. The following December, she was "walking with a cane" and by the fall of 1955 "learning to walk on crutches."[47]Although she never complained about daily injections of ACTH or the restricted diet the drug imposed, she expressed relief when her doctor switched her to an alternative drug regimen. "I am taking the newest wonder drug, Meticorten," she told the Fitzgeralds, "so for the first time in four years don't have to give myself shots or conserve on salt. I feel very fine about that."[48]

In a 1956 letter, Flannery provided one of her first remarks about the lessons to be derived from illness: "I have never been anywhere but sick. In a sense sickness is a place, more instructive than a long trip to Europe, and it's always a place where there's no company, where nobody can follow. Sickness before death is a very appropriate thing and I think those who don't have it miss one of God's mercies."[49] Although loneliness is a common theme in patient narratives, we might ask whether Flannery's presentation of her illness intensified her assumption that she could have "no company." Just as she had tried to conceal her physical distress from the Fitzgeralds in 1950, so she tried to hide all negative emotions when writing to various other friends and acquaintances. As one observer notes, "she rarely chose to take her correspondents into [the] world of suffering. If her life held private terrors she contained them with comedy."[50] She referred to her crutches, for example, as "aluminum legs" and "flying buttresses," and belittled their impact on her life. Using crutches, she insisted, "is not as great an inconvenience for me as it would be for somebody else, as I am not the sporty type. I don't run around or play games. My greatest exertion and pleasure these last years has been throwing the garbage to the chickens and I can still do this, though I am in danger of going with it."[51]

The pressure on Flannery to adopt a cheerful and humorous demeanor must have been especially intense in the 1950s, when the culture demanded that people conceal painful and unpleasant emotions. Some of the least likable characters in Flannery's stories are self-dramatizing invalids. In "Good Country People," for example, Hulga "got up and stumped" her artificial leg, "with about twice the noise that was necessary."[52] Hulga may have embodied Flannery's worst fears about her own behavior. She later confessed to a friend that the character was "a projection of myself."[53] Did the uniformly positive and cheerful image Flannery projected prevent her many friends and acquaintances from sharing her experience? Or did she respond to cues from them that little support would be forthcoming had she more openly expressed her suffering?

In 1957, Flannery did make plans to go to Europe. An aunt offered to send her and her mother to Lourdes in hope of a cure. Although the doctor originally pronounced Flannery too weak to endure such an arduous journey, he finally gave his permission. Sally Fitzgerald, who accompanied the pair on part of the trip, later recounted a long, intimate conversation with Flannery on a train ride. "She said to me . . . that she had come to terms with her illness, with the crippling, the isolation, the constant danger of death, that, in fact, her only remaining fear was that her mother would die before she did. . . . She added, 'I don't know what I would do without her.' "[54] Although Flannery had approached the pilgrimage with her usual mockery and irony, she appears to have harbored at least some hope it would have an effect. Nevertheless, when X-rays demonstrated slight improvement in her hips after her return, she was stunned.[55]

Neither religion nor medicine, however, could effectively retard the onslaught of disease. In 1959 she wrote to a friend that the doctor "says I can't go out of the house without stockings, gloves, long sleeves and large hat. (Sunlight influences lupus and causes joint symptoms etc)."[56] More seriously, she suffered necrosis in her jaws, which made eating difficult and required large doses of aspirin. Public exposure of her illness inflicted a different kind of pain. A *Time* magazine review of her second novel, *The Violent Bear It Away*, included the somewhat inaccurate statement that the author "suffers from lupus (a tuberculous disease of the skin and mucous membranes) that forces her to spent part of her life on crutches."[57] Flannery complained, "That was a really sickening review and in very bad taste." Although her fiction often touched on themes of illness and death, she asserted, "My lupus has no business in literary considerations."[58]

The writings of the French Jesuit priest and philosopher Pierre Teilhard de Chardin helped to sustain Flannery just as her disease was about to enter its final phase. His description of major suffering as a "passive diminishment" struck her as especially insightful. "The 'passive diminishment' is probably a bad translation of something more understandable," she wrote to one friend. "What [Teilhard] means is that in the case he's talking about, the patient is passive in relation to the disease—he's done all he can to get rid of it and can't so he's passive and accepts it."[59] She wrote in another letter, "I think [Teilhard] was a very great man."[60] She expounded on his term while fulfilling a request from the Dominican sisters of Our Lady of Perpetual Help Free Cancer Home in Atlanta. That facility was an offshoot of St. Rose's Home for Incurable Patients, discussed in chapter 4. The sisters asked Flannery to write a novel based on one of their patients, Mary Ann Long, a girl with a malignant facial tumor, who had lived at the home for nine years before her death. Although Flannery refused to tell Mary Ann's story in fictional form, she agreed to write an introduction to the sisters' account of the girl's life. "I am convinced that the child had an outsize cross," Flannery told a friend, "and bore it with what most of us don't have and couldn't muster."[61] The manuscript the sisters eventually produced was "not very good," according to Flannery's assessment. Nevertheless, Farrar, Straus, and Cuday (later Farrar, Straus, and Giroux) published *A Memoir of Mary Ann* in 1961.[62]

Flannery's introduction began by recounting the biography of Rose Hawthorne Lathrop, emphasizing her establishment of the New York home for incurables in

1899. Then, in a passage Flannery may have addressed to herself as well as her readers, she declared that "the creative action of the Christian's life is to prepare his death in Christ. It is a continuous action in which this world's goods are utilized to the fullest, both positive gifts and what Pere Teilhard de Chardin calls 'passive diminishments.' " Although Mary Ann suffered from an "extreme" diminishment, she "was equipped by natural intelligence and by a suitable education, not simply to endure it, but to build upon it." As a result Mary Ann "was an extraordinarily rich little girl."[63]

A few months before the book appeared, Flannery wrote that she was entering Piedmont Hospital in Atlanta to have her "bones inspected." X-rays had revealed more problems in her hips and jaw.[64] From her bed she wrote, "My mother says when we leave, Piedmont will have my money, the doctors will have a lot more information and I will be about where I was when I came in." The physicians' conclusion must have confirmed Regina's skepticism of medicine: ten years of steroid drugs had caused Flannery's bone disintegration. "They are going to try to withdraw the steroids and see if I can get along without them," she reported. If that proved impossible, it would be "better to be alive with joint trouble than dead without it."[65]

In the spring of 1961, Flannery received injections of Novocain and cortisone to alleviate her hip pain, but relief lasted no more than two weeks.[66] She found some basis for hope in the possibility of an operation to insert steel hip joints, recommended by a Providence, Rhode Island, physician, and was disappointed when her Atlanta doctor forbade the procedure. At the beginning of January 1962, she was "on the trail of a new operation. This one they take a piece of bone out of your leg . . . and graft it into your hip bone."[67] As she predicted, however, her doctor considered that surgery equally risky.

"I was interested to hear about the possible lupus," Flannery wrote a correspondent in May 1963. "I am very content with mine. It's latent at the moment and well-behaved."[68] Whatever respite she enjoyed was relatively short. By the summer she experienced overwhelming fatigue, which the doctors diagnosed as severe anemia. Although she received daily iron treatments at a doctor's office, they proved ineffective. "I been sick and am too weak at the moment to care about hitting this typewriter much," she reported at the end of November. "You have to be without energy to gauge whatall it takes."[69] Shortly before Christmas, she fainted and spent the next ten days in bed. Although her internist had rejected previous suggestions of surgeries, he reluctantly agreed that she enter the hospital in February 1964 for the removal a fibroid tumor, now believed responsible for the anemia. "It was all a howling success from their point of view," she wrote to Robert Fitzgerald after returning home, "and one of them is going to write it up for a doctor magazine as you usually don't cut folks with lupus." By the following month, however, it was clear that the surgery had reactivated the disease. Throughout the spring, her letters "continued to pour out" to friends, Sally Fitzgerald later recalled, but "her light tone was deceptive. Most of us didn't realize how sick she was, though in retrospect her letters tell us more than they did at the time."[70] Several ended with requests for prayers.[71]

In early April, Flannery spent several weeks in the hospital, receiving blood transfusions and cortisone injections, but remained desperately ill. By May she was back in the hospital for further tests and blood transfusions.[72] Doctors' reassurances now rang hollow. "They expect me to improve, or so they say," she wrote. "I expect anything that happens."[73] In July she received the sacrament of the sick (previously known as extreme unction).[74] After slipping into a coma, she was rushed to the hospital and died there early in the morning of August 3. The next day she was buried beside her father's grave in a Milledgeville cemetery.

Rachel Carson

During the 1960s, many Americans who fled organized religion forged new ways to retain a sense of the sacred in their lives.[75] Rachel Carson had long succeeded in accomplishing that goal. Although she had renounced her parents' faith years before she was diagnosed with breast cancer at the age of 52, she remained deeply spiritual. Above all, her intense engagement with nature enabled her gradually to come to terms with advancing disease. Because Carson communicated her personal feelings more freely than did O'Connor, we can describe her illness experience in much richer detail.

Four interrelated themes dominating Carson's nature writing had particular salience as she confronted terminal cancer. One was the inevitability of contingency and change. Everywhere she looked, she found nature in flux. The earth's atmosphere was a "place of movement and turbulence, stirred by the movements of gigantic waves."[76] Throughout the history of the earth, "the edge of the sea" had been "an area of unrest where waves have broken heavily against the land, where the tides have pressed forward over the continents, receded, and then returned. For no two successive days [was] the shore line precisely the same."[77] Even the deep sea, long thought to be completely still, was now known to be yet another site of movement and change.[78] The dynamic processes in the world around Carson helped to infuse the transience and unmanageability of her life with meaning. Although she tried desperately to survive, she never doubted that ultimately she would have to face her own end.

Second, the endless cycles of nature meant that birth and renewal continually followed decline and death. "There is something infinitely healing," she wrote, "in the repeated refrains of nature—the assurance that dawn comes after night, and spring after the winter."[79] Feeling the warmth of the sun for the first time in late February, discovering the first crocuses in March, and hearing a spring peeper in April, she was able to align herself with regeneration despite the deterioration she experienced in her body.

Third, nature provided a perspective from which to view all human troubles. As she remarked in a 1951 speech, "When we contemplate the immense age of earth and sea, when we get in the frame of mind where we can speak easily of 'millions' or 'billions' of years, and when we remember the short time that human life has existed on earth, we begin to see that some of the worries and tribulations that concern us are very minor."[80] Especially at a time when illness, suffering, and the routines of care threatened to engulf her life, she relied on her connection to nature to expand the sweep of her vision.

Fourth, because she approached nature with reverence, it provided her with a continual sense of fascination and joy. "If I had influence with the good fairy who is supposed to preside over the christening of all children," she told parents, "I should ask that her gift to each child in the world be a sense of wonder so indestructible that it would last throughout life, as an unfailing antidote against the boredom and disenchantments of later years, the sterile preoccupation with things that are artificial, the alienation from the sources of our strength."[81] Although advancing disease gradually curtailed Carson's ability to tramp though woods, stand at the edge of the shore, and crouch over tide pools, she never lost her capacity for awe.

Carson had been born on May 27, 1907, in Pennsylvania's Allegheny Valley, the third child of Maria and Robert Carson. More inquisitive and tractable than her two older siblings, she quickly became her mother's pride and confidant. Maria carefully nurtured Rachel's interest in nature, taking her on long walks to explore woods and orchards and teaching her to identify the birds and flowers they observed.[82] Rachel also learned from a young age the importance of keeping major parts of her life private. The precarious state of the household income was a source of enormous embarrassment to her, and she worked hard to conceal it from friends.[83]

When her father died suddenly in July 1935, Rachel discovered that her position as the favorite child carried major responsibilities. Because her brother was considered too irresponsible to become the family breadwinner, Maria expected Rachel to support the two of them as well as provide financial assistance to her older sister, the single mother of two daughters. In 1936, Rachel thus joined the federal Bureau of Fisheries (later the U.S. Fish and Wildlife Service). The following year her sister died, leaving Rachel and her mother responsible for two girls, 12-year-old Virginia and 11-year-old Marjie.[84]

Rachel first began writing for the public to earn extra income, publishing *Under the Sea Wind* in 1941 and *The Sea around Us* to widespread acclaim in 1951. Despite her delight in the success of the second volume, she, like O'Connor, hated what she later called "the public's prying curiosity" about her life.[85] And soon she had a new family problem she urgently wanted to keep secret. Her niece Marjie had become pregnant during a relationship with a married man. When Roger Christie was born in September 1952, Rachel announced that Marjie had briefly been married.[86]

Money from *The Sea around Us* enabled Rachel to quit her government position and write full time. In addition, she was able to buy land and build a summer cottage on Southport Island, Maine. During her first summer there, in 1953, she met her neighbor Dorothy Freeman, then 55, married, and the mother of a grown son. Sharing a love of cats and the outdoors, the two women soon declared themselves "kindred spirits."[87] Various scholars have speculated about whether they ever became sexually intimate.[88] What we do know is that, although Dorothy remained married, she quickly became the most significant person in Rachel's life.

Because telephone calls were expensive and family obligations kept both women at home (Dorothy in West Bridgewater, Massachusetts, and Rachel in Silver Spring, Maryland), letters were their central mode of communication.[89] From the beginning, Dorothy worried about the onerous family responsibilities Rachel shouldered. In 1953, Maria Carson was 84 and in frail health. Rachel's niece Marjie lived nearby and suffered from rapidly advancing diabetes, requiring additional help. When Marjie died in January 1957, Rachel adopted Roger, then 5.

Rachel kept her complaints to herself at home but occasionally poured out her frustrations to Dorothy. The "conflict that just tears me apart," she wrote in 1955, was "how to be a writer and at the same time a member of my family."[90] Without a government job, she remained dependent on the income from writing at a time when her financial pressures were rapidly mounting. She increasingly needed to hire domestic helpers and pay medical bills; in 1956, she bought a larger house to accommodate her expanded family. Her next book, *The Edge of the Sea*, garnered rave reviews and prestigious awards after its publication in October 1956 and quickly entered the best-seller list. Rachel then wrote several magazine articles, including "Help Your Child to Wonder." In 1957 she began research for the book that eventually was called *Silent Spring*.[91]

On November 22, 1958, Maria Carson suffered a major stroke. After her death on December 1, Rachel penned this account:

> During that last agonizing night, I sat most of the time by the bed with my hand slipped under the border of the oxygen tent, holding Mamma's. Of course

> I didn't feel she knew, and occasionally I slipped away into the dark living room, to look out of the picture window at the trees and the sky. Sometime between 5:30 and 6:00 I did so. Orion stood in all his glory just above the horizon of our woods, and several other stars blazed more highly than I can remember ever seeing them. Then I went back into the room and at 6:05 she slipped away, her hand in mine. I told Roger about the stars just before Grandma left us, and he said, "Maybe they were the lights of the angels, coming to take her to heaven."[92]

Rachel had insisted that her new house have a view, and she later noted that it had provided her ballast during other sleepless nights with her mother. Although Rachel no longer subscribed to the religious beliefs of her youth, Roger's comment helped to imbue the spectacle with greater significance.

When Rachel sent her editor Paul Brooks drafts of the two chapters of her new manuscript at the end of March 1959, she enclosed a note informing him she was about to go to the hospital for surgery. She hoped it would not be too complicated, although that could not "be known at the moment."[93] As Brooks would later learn, she previously had had breast lumps removed. Now she needed an operation for new ones. If they showed evidence of malignancy, she would undergo a radical mastectomy. This operation, pioneered by Johns Hopkins Hospital surgeon William Stewart Halsted, caused serious chest deformities, including hollow areas under the collarbones and armpits and, in some cases, lymphedema (arm or hand swelling), pain, and mobility problems.[94] Rachel could not know in advance how complicated her surgery would be because her doctor planned to follow the so-called one-step procedure, performing the mastectomy while she still was anesthetized if the pathology report on the biopsied tumor indicated a malignancy. Both the radical mastectomy and the one-step procedure soon would be the object of harsh criticism by doctors and women's health activists alike. In 1960, however, only a handful of physicians challenged the appropriateness of those therapies, and Carson underwent aggressive surgery before waking from the biopsy.[95]

She later explained what happened next: "The tumor was malignant, and there was even at that time evidence that it had metastasized, for some of the lymph nodes also were found to be involved. But I was told none of this, even though I asked directly."[96] Although Rachel's doctors again followed standard practice, she was hardly the typical patient. The research she recently had conducted for *Silent Spring* had given her extensive knowledge about cancer as well as sharpened her skepticism of scientific authority. Some of her friends later said they knew then that the tumor was malignant and that she should receive further treatment.[97] At that point, however, her fierce desire to believe herself well seems to have overwhelmed any more rational misgivings.

When Rachel discovered more lumps in November, her doctors again hedged. She noted that although they admitted the problem "may" have some connection to the former trouble, they "profess to be puzzled."[98] Rachel herself could no longer deny the truth, however. After telling Dorothy about the new development, Rachel apologized: "Just at first I . . . felt I must keep back what was in my mind. And then it all spilled out, and though I felt I should not have done it, I felt so much better."[99] She would often use the word "spill" to express her ambivalence about disclosing disturbing information or emotions to Dorothy. Hearing Dorothy's voice, Rachel had found herself pouring out her news and distress. But she could not avoid feeling that she had allowed something to fall from its proper container.

Others in her situation might have found comfort in the widespread belief that scientists were on the verge of discovering a cure for cancer. Rachel, however, had greeted the arrival of the atomic age with dread and sorrow, and she was all too aware that both radiation and chemotherapy were "two-edged swords," often inflicting dangerous side effects even as they offered the best hope for recovery.[100] Nevertheless, when doctors recommended radiation therapy, she appears to have complied without demur. Informing Dorothy that she soon would begin a course of treatment, Rachel commented that "naturally, there is no choice."[101]

A few weeks later, Rachel realized that she could exercise choice about one issue.[102] After developing flu-like symptoms, she wrote to Dr. George (Barney) Crile, Jr., an iconoclastic Cleveland surgeon whom she had met socially. Although best known for his condemnation of the widespread use of radical mastectomies, Crile also espoused a general approach to patient care sharply at odds with that of the rest of the medical profession. "I had admired his little book on cancer greatly when it appeared," Rachel told her editor, "and I had thought then that if ever I had such a diagnosis I would want to consult him."[103]

Crile's 1955 *Cancer and Common Sense* opens with a description of a seriously ill woman sitting on a Corsican hill close "by the tomb of her ancestors, looking at the sea." Crile could only hypothesize that she suffered from cancer, but he did "know that this black-robed woman felt herself to be one with her ancestors and with the sea. There was no fear in her face. She knew that her death was a part of her life, a part of the stream of life, a part of the sunshine, of the flowers, and of the blueness of the sea."[104] Although we may question the extent to which Crile could "know" the intimate feelings of a woman who lived in a very different culture and to whom he never spoke, we can well imagine that the scene he painted would have a profound impact on Rachel Carson, who had long felt connected to the oceans and had recently published a trilogy about them.

Rereading *Cancer and Common Sense* in 1960, she must have been struck by its similarity to the manuscript she was preparing. The primary message of *Silent Spring* is that humans must approach the natural world with humility and awe, respect its "integrity," and exercise enormous caution before tampering with it. The concluding paragraph begins: "The 'control of nature,' is a phrase conceived in arrogance, born of the Neanderthal age of biology and philosophy."[105] Crile displayed equal contempt for medical hubris. Writing at a time when some surgeons advocated "super radical mastectomies," he directed his greatest wrath toward those procedures. But he noted that even radiation could harm surrounding tissues. And many treatments were unnecessary because people could survive "long and comfortably" even with advanced disease. Rather than promising to avert death through ever more aggressive therapy, doctors could teach patients "to live with their cancers with equanimity."[106]

After flying to Cleveland to consult Crile in mid-December, Rachel thanked him for having enough respect for her "mentality and emotional stability" to talk frankly with her. "I have a great deal of peace of mind when I feel I know the facts, even though I might wish they were different."[107] Although one of those facts must have been that her cancer had spread, we can assume that Crile reiterated the message in *Cancer and Common Sense*, that metastasized disease did not always mean a bad outcome: "Each cancer follows its own pattern of growth; each has its own peculiarities of behavior; each runs its own course."[108] Rachel's correspondence provides other clues that Crile encouraged optimism. "There is much that is hopeful in my case," she wrote to Brooks, "and certainly now that I've had a chance to make some decisions for myself, I'm in excellent hands."[109] Dorothy must have received similar reassurance because she expressed "great sense of relief" after hearing about Rachel's visit to Cleveland."[110] But Crile also must have told Rachel plainly that her life might well end prematurely. She later wrote about the deep sense of solemnity she felt on the trip home; she understood that her unfinished tasks had acquired a new urgency and that she must clarify her goals and savor the present. At Christmas she expressed gratitude that she and Dorothy had "learned to cherish every such moment."[111]

Rachel now had yet another major part of her personal life she needed to hide. Soon after Christmas, she told Paul Brooks that she planned to inform her agent about recent developments but then added: "As far as other people are concerned . . . I don't want to have it discussed at all. Any such facts about a writer are immediately spread about in such a way that is most distasteful to me."[112] Carson's overriding concern, however, was not hiding the diagnosis from the public but rather communicating her illness experience to the woman who shared her life. Dorothy's visit early in January 1961 enabled both women to exchange the feelings they hesitated to express in letters. "Best of all," Rachel exulted after Dorothy's departure, "was the chance to share each other's thoughts and in their light to exorcise any lurking dark spirits. I want you to think only happy thoughts of me from now on, darling, and now that we have faced this problem together I hope and believe you can."[113]

Those happy thoughts could involve more than optimism about recovery. In a January 1961 letter, Rachel enclosed a photograph of Crile with his wife Jane, who recently had revealed that she too suffered from breast cancer. "I think even her picture will tell you she is so vibrantly and joyously alive," Rachel wrote. "Now to know she, too, has encountered this shadow and in spirit at least has triumphantly overcome it is a wonderful example."[114]

A cascade of serious medical problems soon overwhelmed Rachel. Crile had recommended a new course of radiation, first to her ovaries and then to her left chest. Shortly after treatment began, she developed a staphylococcus infection in her bladder and recalled that a similar affliction had claimed her niece Marjie four years earlier. One afternoon Rachel was "so indescribably weak and ill" that she felt as if her "life had burned down to a very tiny flame, that might easily flicker."[115] Severe joint inflammation in her legs lasted longer, caused even greater suffering, and forced her to enter the hospital, where she received physical therapy as well as radiation.[116]

Back home at the end of a week, she reveled in her growing independence. Although the interconnectedness of all forms of life was a major theme of her writing, Carson shared the cultural glorification of self-sufficiency and hated the dependence sickness necessitated. It was thus a tremendous relief to take "the first step toward shaking off" some of the helpers she had hired to care for herself and Roger.[117] Nevertheless, life remained immersed in illness. Although she quickly abandoned the physical therapy, trips to the hospital for radiation consumed most days. Because she could neither walk nor drive, she needed various types of assistance; each journey was "quite a safari," requiring "car, wheel chair, and assorted bearers and pushers."[118] Despite her medical insurance, other expenses mounted; by mid-February, her money was "pouring out at the rate of literally hundreds of dollars a week." Some treatments were painful. And all produced "miserable" nausea, making work impossible.[119]

The many misfortunes that arose during the first part of 1961 intensified Rachel's dilemma of communicating openly with Dorothy without igniting her fears. Dorothy encouraged Rachel to voice her fears and sorrows freely, but after rereading another "very cheerless" letter she had just finished writing, Rachel hastened to add that it "does not reflect my general mood." She was "just spilling out" the "occasional thoughts of the dark hours."[120]

The primary way Rachel managed to keep her accounts of medical horrors tolerable was not by censoring herself or minimizing her complaints but by detailing the joy she continued to experience in nature. After describing the ambulance trip to the hospital and the "awful" "bed-pan service," she noted, "From my 5th floor window I look west toward McMillan Reservoir and—guess what?—almost every time I look out I see one or more gulls sailing over it." She wished she had brought her field glasses and bird guide.[121]

By late February she was able to chronicle spring's arrival. One night she "became aware of a high, sweet, bubbling call coming from the Wildwood. The year's first frogs!" The following morning she heard song sparrows and mourning doves. Early in March Rachel spotted "golden crocuses blooming in someone's yard" and a robin in her own. Soon bluebells were coming up and the first grackles were "glistening the sun." "All these reminders that the cycles and rhythms of nature are still at work are so satisfying," she concluded.[122]

Dorothy enabled Rachel to experience many sights she could not witness herself. "What a treat the mail bought!" Rachel exclaimed after finding a detailed description of a day at the beach in late February. "It really is the very next thing to being there to have read your vivid account."[123] Dorothy's letters could not stifle all regrets. Thinking about her "daffodils coming into bloom" during a particularly difficult day in March, Rachel longed "to walk out and take inventory!" "Each spring is so precious," she noted. "I wanted this one, and feel cheated."[124] Far more often, however, she delighted in the signs of regeneration and renewal, which helped to counter her own sense of loss and decline.

With the end of radiation in March, Rachel's nausea receded. In April, her mobility gradually began to improve, and by June she was well enough to travel to Maine for the summer. Unfortunately, few letters survive from the late spring, summer, or fall, when she wrote several chapters of *Silent Spring* and revised those already drafted. An eye inflammation developed in November, threatening to delay the work still further. "Yes," she exclaimed in January 1962, "there is quite a story behind *Silent Spring*. Such a catalogue of illnesses!"[125]

To what extent did Rachel tell that story in the book? Perhaps because the chemical industry later claimed that her cancer diagnosis caused her to exaggerate the dangers of pesticides, historians have scrupulously avoided that question. She had selected the topic, conducted the research, and written several chapters long before she knew she had breast cancer. Nevertheless, she had ample opportunity to express her feelings about her medical troubles once she returned to the manuscript in the spring of 1961. When she complained that public protests against pesticides met "little tranquilizing pills of half truth" and "false assurances" from industry and government officials, was she remembering the misinformation she received after her mastectomy?[126] Did she contemplate the damage radiation may have done to her body when she wrote, "Along with the possibility of the extinction of mankind by nuclear war, the central problem of our age has . . . become the contamination of man's total environment with [chemicals] of incredible potential for harm"?[127] And what did it mean to her to write a book about the possibility of devastating loss at a time of heightened awareness about the fragility of her own life? Although she never used the word "death" when writing to friends, it studded the book. One chapter was entitled "Elixirs of Death," another "Rivers of Death." The tone of *Silent Spring* also differed dramatically from that of Carson's personal communication. As Dorothy often remarked, Rachel rarely exposed her anger in private. *Silent Spring*, however, is filled with rage. Outlining her goals to her editor in 1958, she noted that she intended to "reveal the futility and the basic wrongness of the present chemical program" by describing it rather than "ranting against it, though doubtless," she added, "I shall rant a little, too."[128] Was Rachel able to discuss death and express anger in *Silent Spring* because she viewed the book solely as a professional project, divorced from her personal life? Or is it possible that in railing against the destruction wrought by pesticides she also raged against the prospect of her own mortality?

If we can only speculate about the answers to such questions, it seems clear that her uncertainty about her own survival intensified her determination to preserve the basic fabric of the natural world which the chemical industry threatened to destroy. Early in January 1962, she recalled the plane trip home from Cleveland the previous year: "I knew then that if my time was to be limited, the thing I wanted above all else was to finish the book."[129] In the middle of the month, she sent fifteen chapters to both her agent and William Shawn, who had agreed to publish excerpts of the book in the *New Yorker*. When Shawn phoned with enthusiastic praise, she realized she had fulfilled her goals, realized her ideals, and created her legacy.[130]

But she had little time to savor her triumph. In February she discovered hard lumps in her right lymph nodes, and her doctors recommended surgery. At the beginning of March she flew to Cleveland to get another opinion from Crile, who confirmed that she now had extensive metastasis on her right side but advised radiation rather than surgery. Shortly after her return, she thus embarked on yet another round.

"Another milestone," she reported in mid-March. After "one more" treatment, the first week would be over.[131] Soon she was again expressing discouragement and then apologizing for the revelation. "I hope you weren't sad after our conversation," her March 26 letter to Dorothy read. "I shouldn't have let you see my courage had deserted me for a time. But it's all right now—all the 'fight' is back in me today."[132] Part of her renewed boldness stemmed from a conversation with Dr. Morton Biskind. Rachel had relied on Biskind's research in *Silent Spring.* Now he recommended alternative medical remedies to help her manage the side effects of radiation.[133]

The regeneration of nature further buoyed her spirits. "Last night's rain has suddenly brought the lawn to life in a surge of green," she wrote on April 1. "And everywhere there is a mist of green from the willows and of red from the maples. The blue buds that I first took to be grape hyacinth opened into my lovely Chinodoxa." As before, some signs of spring were bittersweet: "Recently I realized with a start that it was too late to plant freesias—and I had so wanted them again."[134] Nevertheless, curiosity and awe continued to trump regrets.

Rachel would have even more reason to turn to those sights in the weeks ahead. "Well," she wrote on April 10, "the news is somewhere between my hopes and fears—better than my fears, or some of them, but worse than my hopes."[135] She had experienced more pain and soreness in her armpit and thought she felt something. Tests revealed "another enlarged node," indicating the radiation had not prevented continued metastasis. The good news was that the neck pain was "just the effect of treatment" and the soreness in her back resulted from "arthritis quite consistent with my age." But even benign symptoms served as a reminder that watchful waiting could never end: "The trouble with this business it that every perfectly ordinary little ailment looks like a hobgoblin, and one lives in a little private hell until the thing is examined and found to be nothing much." Fortunately, she could provide a nature report as well as a medical one: "This has been a beautiful day of blue, blue skies, all green and gold with willows and forsythia."[136]

In May 1962 Carson reiterated her determination to keep her diagnosis secret. Dorothy should tell mutual friends that she "*never saw*" her "*look better.*" Rachel had new awareness of "what happens when even an inkling of the other situation gets out." The previous evening she had overheard "scraps of dinner conversation about poor Senator Neuberger: 'You know she had a cancer operation.' . . . 'They say she's down to 85 pounds.' . . . 'If you'd see her on the Senate floor you'd know she can't last.' " That talk was "just the sort of thing" Rachel "couldn't bear." "Whispers about a private individual might not go far," she emphasized. "About an author-in-the-news they go like wildfire."[137] And soon there were many indications that her fame was about to soar. "What I never counted on—really never even allowed myself to hope for—has happened," she jubilantly wrote on June 11. "*Silent Spring* will be the October Book-of-the-Month."[138] The first installment of the *New Yorker* series appeared five days later. By the end of the month she knew that after its September publication, the book would be a "smashing success."[139]

As Rachel contended with both her rising renown and the barrage of virulent criticism unleashed by the chemical industry throughout the fall, the specter of worsening health continued to haunt her.[140] After describing an October reception in her honor in Cleveland, she commented, "All this written as though the menacing shadow did not exist, yet the day before I left it seemed as though time was standing still and there might even be no tomorrow."[141] By the end of the year, that shadow had darkened considerably. The "main thing" concerned the back pain. The radiologist had reexamined X-rays taken early in December and decided that "although nothing shows, he felt it advisable to use some radiation."[142] As Carson later indicated to Crile, she realized that when the radiologist said she did "not necessarily" have a metastasis in the spine, he was "just trying to reassure me."[143] Nevertheless, she used the same equivocation in her letter to Dorothy: "The diagnosis is by no means definite, and it is possible the trouble has no relation to the malignancy." In addition, she noted, "if it really is a metastasis, then we have gotten it quite early, before any real damage has been done. So there is much reason for optimism."[144]

Rachel also tried to belittle the second problem. A week before Christmas, she had fainted while buying a present for Roger. Although a cardiogram revealed "nothing much," both Crile and her radiologist urged her to consult a heart specialist because she also had experienced recurrent chest pain. "Well," she concluded, "I may do something about it one of these days, but if it is angina certainly it's a mild case and I'm not running up any hills or anything like that!"[145]

But she could not easily trivialize that complaint in practice. A few weeks later the pains were so frequent and intense, that she feared death was imminent and decided to write a good-bye letter. Rather than mailing it, Rachel left it in an envelope addressed to Dorothy. The letter reflected first on the way they previously had handled painful separations: "When I think back to the many farewells that have marked the decade . . . of our friendship, I realize they have been almost inarticulate. I remember chiefly the great welling up of thoughts that somehow didn't get put into words—the silences heavy with things unsaid. But then, we knew or hoped, there was always to be another chance—and always the letters to fill the gaps." Soon, however, the time might come when there could be no more letters. Rachel thus had to seize the opportunity to say what previously had been impossible. She hoped Dorothy would have no regrets on Rachel's behalf. "I have had a rich life, full of rewards and satisfactions that come to few, and if it must end now, I can feel that I have achieved most of what I wished to do. That wouldn't have been true two years ago, when I first realized my time was short, and I am so grateful to have had this extra time."[146]

Despite gratitude for that time, Rachel clearly shared the prevailing wisdom that a sudden cardiovascular death was preferable to a lingering cancer one. She added that although the angina was a cruel blow, it felt "almost like a secret weapon against the grimmer foe." If it "should take" her quickly, that would be "the easiest way." The letter concluded with thoughts she would repeat over and over in the coming months: "I want to write of . . . the joy and fun and gladness we have shared—for these are the things I want you to remember—I want to live on in your memories."[147]

Rachel's regular correspondence continued to chronicle the daily experience of a life enmeshed in illness. The cardiologist she saw at the end of January imposed a new set of restrictions, forcing her to become "a prisoner again."[148] She could not climb stairs or even leave the house, and she slept in a hospital bed with her head elevated.[149] The return of the chest pains in the middle of February was thus discouraging: "Short of completely giving in to an invalid's life, I scarcely know what less I could do." Because she had told "almost no one" about the angina, she remained lonely and isolated: "Everyone just thinks I'm still rushing around being busy—so there are no get well cards, inquiring calls, or any of the things that perhaps make sickness a little less solitary affair."[150] Like Flannery O'Connor, Rachel had discovered that the concealment that protected her public image prevented her from soliciting the sympathy and support she craved.

As her cancer proceeded on its downward course, she also continued to negotiate how openly to communicate to Dorothy. After informing her in an evening phone call about the discovery of new tumors in her lymph nodes, Rachel instructed Dorothy how to frame this latest setback: "Remember, darling, it is just another of the series of battles I knew I must face, and I am sure it is not the last. We will win this one. You know the lymph tumors are very susceptible to radiation."[151] But Rachel seems to have had no illusion that she ultimately would vanquish the foe. Even while assuring Dorothy she would "win this one," she implicitly acknowledged that the outcome of subsequent battles might be very different.

Carson was far more direct in a letter to Crile, reporting what she had failed to reveal to Dorothy—evidence that the cancer had spread to her bones as well as her lymph nodes. "Doesn't all this mean the disease has moved into a new phase and will now move more rapidly to its conclusion?" she asked. "I still believe in the old Churchillian determination to fight each battle as it comes. ('We will fight on the beaches'—etc.) But still a certain amount of realism is indicated, too, So I need your honest appraisal of where I stand."[152]

Crile responded immediately by phone, reassuring Rachel that the radiation might prove as effective with this new metastasis as it had with the old. Two days later, she wrote to Dorothy: "There is one more medical fact I must give you." Recent X-rays had shown conclusively that her shoulder pain came from cancer, not arthritis as she had hoped. She repeated Crile's hopeful message about radiation. And she reminded Dorothy that their heightened awareness of the transience of life need not diminish the joys they shared.[153]

Nevertheless, her March reports were grim. Her back pain remained severe despite the radiation, and the treatments produced fatigue and "troublesome" nausea.[154] That month she wrote the first of two remarkable letters about the meaning of her approaching death. Dorothy had introduced the topic of immortality in a letter on March 6, and three weeks later Rachel finally felt well enough to respond. Like most of her most important missives, this one began with a description of nature. That morning she had "uttered a loud 'Oh!'" when she opened the blinds in the window of her study: "One large clump of daffodils is suddenly in full bloom—there must be 6 or 8 blossoms. Well, spring must be just about on schedule."[155]

Her deep connection to and understanding of nature also informed her musings on immortality. She first referred to the conclusion of her 1957 article "Undersea," which read: "Individual elements are lost to view, only to reappear again and again in different incarnations in a kind of immaterial immortality. . . . Against this cosmic background the life span of a particular plant or animal appears not as a drama complete in itself, but only as a brief interlude in a panorama of endless change." Although that thought still had "great meaning and beauty," she realized it was "purely a biologist's philosophy." The notion of "immortality through memory" seemed "far more satisfying." It was a source of great pleasure for Rachel to know that she would "live on even in the minds of many who do not know me, and largely through association with things that are beautiful and lovely." And, of course, regardless of whether she or Dorothy died first, they would continue to speak through "many things—the songs of the veeries and hermits, and a sleepy white throat at midnight—moonlight on the bay-ribbons of waterfowl in the sky."

Even that idea, however, was not sufficient. "How could that which is truly one's self cease to exist?" she asked. "And if not, then what kind of spiritual existence can there be?" Although any type of life after death lay beyond human comprehension, she could still believe in it. Here she drew on another of her famous nature articles. The end of "Help Your Child to Wonder" had described a Swedish oceanographer who had told his son that as he faced death, he would be sustained "by an infinite curiosity as to what was to follow." To Rachel, with her insatiable inquisitiveness about the natural world, that "sort of feeling," served as "an acceptable substitute for the old-fashioned 'certainties' as to heaven and what it must be like." She was glad Dorothy had broached this topic and hoped they would discuss it again. After all, it was "not a gloomy subject" and did not relate only to her own situation, "for this is something we all share—a normal part of life." Recalling that Crile had compared "the life-death relationship to rivers flowing into the sea," Rachel concluded that the analogy was "not only beautiful but somehow a source of great comfort and strength."[156]

Dorothy replied that Rachel's letter had "opened up new vistas for my mind to travel along."[157] Rachel's own life, however, continued to contract. On April 7, she was "almost glad" Dorothy could not see her. Although she could endure the pain, the "crippling, real and prospective" was far more difficult. She dreaded each morning, fearing that her mobility would have diminished since the previous day. "I shouldn't be saying all this," she added, but Dorothy was planning to visit, and Rachel wanted to prepare her.[158]

The one hope lay in a popular alternative remedy. First introduced by a Yugoslavian doctor and his brother in 1949, the anticancer drug Krebiozen soon received the support of a former chair of the National Cancer Institute's National Advisory Council on Cancer. But the drug also aroused virulent criticism from prominent physicians, some of whom derided the substance as a combination of horsemeat and mineral oil. In 1962 the government banned the interstate commerce of the drug.[159] During the spring of the following year, when Rachel tried to convince various doctors to administer the treatment, the Federal Drug Administration was investigating it in the wake of the thalidomide tragedy.[160]

From Rachel's vantage point, Krebiozen offered significant advantages. It was a systemic rather than a local treatment, helping "the whole body resist." Because it was nontoxic, she might be spared the terrible side effects of radiation. She insisted that the controversy surrounding the therapy reflected "the bickerings, struggle for power, bigotry, etc., within the medical profession rather than any valid objection to the drug."[161] Indeed, she told Crile, the attacks by the American Medical Association "resemble so closely some of the methods used against those critical of pesticides that the parallel is quite suggestive."[162] Rachel began the treatment in early April. On May 2, she conceded that the pain had not abated. Krebiozen was "still a hope, but only that."[163] She abandoned it the following month.

Rachel had warned Dorothy that she might not be able to return to Maine for the summer, but she finally arrived at her beloved cottage on June 25, accompanied by a friend to watch over her. One day in early September, Dorothy drove Rachel to one of their favorite places, the Inn at Newagen, from which they could see Todd Point and Griffiths Head. Later that afternoon, Rachel wrote her second extraordinary letter about the relationship between nature and the meaning of death. "This is a postscript to our morning at Newagen," she began, words that were easier to write than to speak.

> Most of all I shall remember the Monarchs, that unhurried westward drift of one small winged form after another, each drawn by some invisible force. We talked a little about their migration, their life history. Did they return? We thought not; for most, at least, this was the closing journey of their lives.
>
> But it occurred to me this afternoon, remember, that it had been a happy spectacle, that we had felt no sadness when we spoke of the fact that there would be no return. And rightly—for when any living thing has come to the end of its life cycle we accept that end as natural.
>
> For the Monarch, that cycle is measured in a known span of months. For ourselves, the measure is something else, the span of which we cannot know. But the thought is the same: when that intangible cycle has run its course it is a natural and not unhappy thing that a life comes to its end.[164]

All summer Rachel's mobility had steadily diminished, and on the day of her departure, she could barely walk to the car. As she soon learned, the cancer had spread to her pubic area and a bone had fractured. Her doctors recommended a phosphorus hormonal therapy, which she began at the end of September. "This *has* to work," she stressed, reminding us that acceptance rarely is achieved once and for all.[165] Each new remedy revived hope; each new sign of advancing cancer forced her to learn the lesson of the monarchs yet again.

Although Rachel had doubted she would be able to travel to San Francisco to give a major address, her doctors remained optimistic, and by the middle of October she had improved enough to be able to fly across the country, accompanied by her agent. She used a wheelchair whenever possible, relied on a cane to get on and off the stage, and sat to deliver her hour-long speech, explaining to anyone who asked that she suffered from arthritis.[166]

Back home, she realized she was even more limited than before. Because disease had spread to her upper back and arms, she was in too much pain to use the walker; numbness in her right arm made writing difficult.[167] Three new blows fell between November 1963 and January 1964. The first was President Kennedy's assassination. Rachel had greatly admired both the man and his administration, and he had staunchly championed *Silent Spring.* Too distraught to sleep during the "awful" nights that followed, she searched her bookcase for something to read. Nothing appealed until she saw *Under the Sea Wind.* From then on, she found that a "chapter or two" in bed relaxed her enough to sleep. "Of course," she explained, "it is the elemental nature of the subject matter, its timelessness, beside which human problems and even human tragedy fall into perspective." For the first time she understood why so many people turned to her books about the sea "in time of trouble."[168]

The death of her favorite cat in December represented another major loss. "So many sad and somber thoughts," Rachel wrote the following morning. "For exactly three years, since I flew to Cleveland in December and first understood my situation, I have worried about my little family." She knew that whoever adopted Roger would not want an animal as well. When the other cat had died at end of the summer, Rachel had felt that the "inevitable dissolution" of her "little circle had begun." Now she had "lived to witness another step."[169]

Then, in January, Dorothy's husband, Stan, died suddenly of a heart attack while he sat at his kitchen table watching birds. Rachel hoped that, even in the midst of her anguish, Dorothy understood that the death was "wonderful for him." He had "no apprehension, no pain, just sudden oblivion shutting down while he was in the midst of one of his happiest occupations." Echoing Rachel's awareness that a cardiovascular death might be her "secret weapon," Dorothy wrote, "Certainly it is what we would all choose if we had to."[170]

Rachel's own troubles continued to mount. As her neurological problems advanced, she lost the sense of smell and taste. At the end of January, she developed a lung infection requiring hospitalization, followed by uncontrollable nausea, meningitis, shingles, and anemia. Crile recommended that she receive an operation to ablate her pituitary gland by implanting a radioactive isotope. Having exhausted all other remedies, she finally consented in March, flying to a Cleveland clinic where he could supervise her care. Such a desperate measure might seem totally out of character for Rachel as well as her doctor. She had repeatedly expressed her wish for a peaceful end. Crile had often railed against the use of aggressive actions for persons who were gravely ill. And both had preached the need to accept finality. And yet, in the event, neither could refuse one final attempt to avert death. Rather than emphasizing her hope for a reprieve, Rachel asked Dorothy to "remember the joys we have shared, the love each has felt for the other; all this is enough for a lifetime."[171] Although Rachel survived the operation, she remained critically ill, experiencing heart irregularities as well as jaundice from liver metastases. She flew home on April 6 and died eight days later—shortly before her 57th birthday.

At a time when medicine promised to find a cure for virtually every malady and popular culture encouraged patients to fight rather than prepare for mortality, terminally ill people may have been especially likely to remain unreconciled to the "dying of the light." Nevertheless, some continued to turn to religion to make sense of their experiences. As the writer Karen Armstrong notes, "All the world's faiths put suffering at the top of their agendas."[172] Although the patients examined in this chapter desperately wanted to survive, took advantage of every available treatment, and mourned the losses sickness inflicted, both they and their relatives gradually accepted suffering and death as inevitable aspects of life. "This is only our human condition—common to all," Dorothy Dushkin wrote shortly before Amanda died. Dorothy could thus "yield to this suffering" which has come "just as clouds come over the mountain." Flannery O'Connor found insight in the suffering she endured, writing to a friend that "sickness is a place, more instructive than a long trip to Europe." Faith traditions also allowed both patients and family members to address ultimate issues. "The impending death of one's child," Frances Gunther wrote, "raises many questions in one's mind," including "What is the meaning of life?" and "What are the relations between Man, men, and God." And several people were able to place personal tragedies in a larger context. Frances concluded that Johnny's death was "only the final scene in a single act of a play that goes on forever."

Rachel Carson used her deep bond with nature in similar ways. When her infirmities kept her confined to the hospital or a "prisoner" at home, she relied on the views from her windows to remind her of her part in a larger whole. As her disease followed a relentlessly downward trajectory, she found satisfaction each spring in knowing that "the cycles and rhythms of nature" were still at work. And despite overwhelming physical suffering, she repeatedly turned to her favorite flowers, trees, and birds for a sense of wonder and awe.

Conclusion

Although death is a universal human experience, the care of dying people traditionally has occupied a marginal place in medical history. Focusing on that topic reconfigures the field. This book demonstrates that most of the benefits from the transformation of medicine and health care at the turn of the twentieth century bypassed individuals who were terminally ill. Medicine's mandate to avert death increasingly took priority over the duty to relieve pain and suffering at the end of life. Celebrating their new healing powers, physicians routinely shunned patients with little hope of recovery. Hospitals restricted the entry of people with hopeless prognoses and quickly discharged those who gained admission.

The exclusionary policies adopted by hospitals meant that large numbers of people ended their days in other medical institutions. Numerous studies of tuberculosis sanatoriums exist, but medical historians have largely ignored chronic disease hospitals, homes for incurables, almshouses, and old age homes. Those institutions both reflected and reinforced the negative attitudes surrounding patients with hopeless prognoses during the late nineteenth and early twentieth centuries. Low funding kept the conditions of public facilities at dismal levels. Staff and administrators in both public and private institutions openly expressed their revulsion at the visible manifestations of advanced disease. Some doctors and nurses tried to reorient their institutions to the care of patients with more favorable prognoses. And even facilities with extremely high mortality rates issued reports concentrating on the rare patient discharged; virtually none reflected on the importance and meaning of the lives of patients approaching death.

Families and religion have also figured prominently in this book. Despite the establishment of various institutions housing dying individuals, much caregiving remained at home. And contrary to Ariès's claim that families welcomed institutional placement as a way to distance themselves from the dying, most relatives were loath to sever ties with people at the end of life. Especially as death drew near, affluent families took advantage of the liberal visiting hours of private rooms to hold around-the-clock vigils. Others made enormous sacrifices to remain close to the dying relative, using scarce resources to pay for private rooms, traveling long distances, and temporarily relinquishing employment and domestic responsibilities. When those options were unavailable, the knowledge that a loved one had died alone compounded the grief of the bereaved.

Although the spiritual components of the "good death" had largely disappeared by the post–World War II period, some patients and their families continued to look to religion to frame the experiences of suffering that medical science could not prevent. Dorothy Dushkin drew on an American brand of Buddhism to remain immersed in the present as her daughter's life slowly ebbed. After her son's death, Frances Gunther sought solace from her "spontaneous, intuitive sense of the reality of God." And using very different spiritual approaches, Flannery O'Connor and Rachel Carson were able to imbue their impending deaths with meaning. Placing death and dying at the center of the discussion thus reminds us that medical history extends far beyond hospitals, doctors, and nurses.

This book concludes just before the 1965 passage of Medicare and Medicaid, programs that shifted much of the cost of the care for those with terminal illnesses to the government, alleviating some of the most terrible access problems I have chronicled. Three-fourths of dying patients are 65 and older and eligible for Medicare. Some also qualify for Medicaid.[1] Thirty-two percent of Medicare expenditures go for the care of people in the last two years of life.[2] Despite the widespread belief that Medicaid overwhelmingly serves poor mothers and children, it is the primary funding source for long-term care; a very high proportion of recipients suffer from fatal chronic conditions.[3]

Two other developments after 1965 affected mortality in contradictory ways. Arguing that the technological, curative focus of modern medicine had distorted the process of death and denied its inevitability, the hospice movement sought to increase acceptance of the course of nature, relieve unnecessary suffering, and "restore dignity to the dying."[4] Simultaneously, the use of expensive, new life-prolonging technologies rapidly increased. As a result, many dying people receive extremely aggressive hospital treatment, in some cases against their wishes. The history examined in this book can help us understand both the widespread appeal of the hospice movement and the ascendancy of high-intensity hospital services at the end of life.

By the late 1950s, a growing number of commentators had begun to call for a movement to humanize the care of dying people. Physicians expressed concerns about the use of heroic treatments when death was imminent. "There are too many instances," wrote Edward H. Rynearson, M.D., in which patients in extremis and "suffering excruciating pain" are "kept alive indefinitely by means of tubes inserted into their stomachs, or into their veins, or into their bladders, or into their rectums—and the whole sad scene thus created is encompassed within a cocoon of oxygen which is the next thing to a shroud."[5] Others questioned practices of concealment, noting that patients often intuited their diagnoses and that secrecy exacerbated their sense of loneliness and isolation.[6]

In *The Meaning of Death*, published in 1959, Henry Feifel wrote that "in the presence of death, Western culture, by and large, has tended to run, hide, and seek refuge in group norms and actuarial statistics. . . . Concern about death has been relegated to the tabooed territory heretofore occupied by diseases like tuberculosis and cancer and the topic of sex."[7] Essays in this volume by various psychologists, chaplains, and humanities scholars attempted to counter that taboo. The sociological studies I have cited by Barney G. Glaser, Anselm L. Strauss, Jeanne C. Quint, and David Sudnow represented another milestone. Published in the mid- and late 1960s, those works demonstrated the results of medicine's overwhelming focus on averting mortality. People dying in both public hospitals and the wards of private ones were especially likely to suffer abuse and neglect. Elisabeth Kübler-Ross's 1969 book *On Death and Dying*, based on interviews with 500 terminally ill hospital patients, attracted a far more popular audience.[8] Although her theory about the five stages of grief (denial, anger, bargaining, depression, and acceptance) increasingly provoked criticism, she marshaled impressive evidence to demonstrate that hospitals overtreated terminally ill patients, isolated them from their families, and abandoned them when they were most needy.[9] By 1976, the book had sold more than a million copies.[10]

Another pivotal event in efforts to humanize death was the 1963 trip of British doctor Cicely Saunders to the United States. A leading critic of hospital care for the dying, Saunders chastised doctors who concluded there was "nothing more to be done" for patients desperately needing pain relief and spiritual and emotional solace.[11] A major purpose of her trip was to visit American institutions for the terminally ill, including the two facilities established by Rose Hawthorne Lathrop, which Saunders pronounced models of compassionate care. In addition, she lectured widely about her research on pain control for the dying and her vision for St. Christopher's in the Field, the hospice she soon would establish in England. At Yale University, her audience included the dean of the School of Nursing, Florence Wald, who later remarked: "When I heard [Saunders], that just opened the door to me. It solved the problem that both the faculty and the students were having in the hospital, seeing patients, particularly cancer patients, being treated with curative treatment, and where it was very obviously not curing the disease, but the suffering was so great. . . . They couldn't get the doctors to tell them what the . . . outlook was for them, or to consider a variety of ways of treating the situation."[12]

In 1966, Saunders returned to Yale as a visiting professor at Wald's invitation.[13] Two years later, Wald took a sabbatical at St. Christopher's, now in operation in London. Soon after her return, she resigned her position as dean to devote herself to the establishment of the first U.S. hospice. The Connecticut Hospice Institute opened in Branford, Connecticut, in 1974. Seven years later, 800 other hospices either existed or were being planned.[14] The creation of the Medicare hospice benefit in the 1982 Tax Equity and Fiscal Responsibility Act both demonstrated the widespread popular support for the hospice idea and facilitated further growth. In 2007, approximately 1.4 million people, representing nearly 40 percent of all deaths, received hospice care; in many cases, hospice services enabled patients to die at home, often the preferred site of death.[15] Many other patients have benefited from the palliative care programs, incorporating many elements of the hospice philosophy, established in a growing number of hospitals and nursing homes.[16]

One explanation of the rapid embrace of the hospice ideal is that the movement arose during a period of widespread social reform. As Wald reminded an interviewer, "It was the same month, almost the same day, when I heard Cicely Saunders, that Martin Luther King had the first march in Selma, Alabama. . . . It was a time of much more open criticism." Protest also reached medicine.[17] Hospices may have seemed especially appealing because they incorporated the ideas of diverse health care reform movements. The holistic health movement demanded that patients be viewed as entire human beings, not simply the sum of their organs. The women's health movement challenged the sovereignty of physicians and the technological focus of modern medicine; a branch, the home birth movement, criticized the medicalization of pivotal life events. And the proponents of self-care advocated the demystification of medical knowledge as well as the promotion of more equal relationships between patients and physicians.

An even more important reason for the rapid proliferation of hospices is that they addressed major, long-standing critiques of hospital care for the terminally ill. A basic premise of the hospice philosophy was that the entire family was the unit of care. Leaders of the movement castigated hospital regulations that disrupted intimate ties just when the need for them was greatest. Although death was a solitary event, hospice advocates insisted it must not be lonely. All rooms at the Connecticut Hospice were "large enough to permit family members of each patient to be there."[18] Many other programs provided services to people dying at home. And all sought to assist relatives as well as patients. Recognizing that death was a trauma for family members, hospices encouraged them to express feelings of anticipatory grief and furnished bereavement counseling to them after the death. Many hospices also offered special services to relieve the burdens on family caregivers.[19]

A second central tenet of the hospice movement was that dying people had emotional and spiritual needs, not just physical ones. As we saw in chapter 3, a few early-twentieth-century reformers established social service departments and upgraded chaplaincy programs in response to the growing depersonalization of hospital care. But social workers and clergy remained marginal players in medical institutions. In hospices, they were considered equal members of interdisciplinary teams.[20] Spiritual care was especially critical to hospice founders.[21] Florence Wald, for example, worked closely with clergy from diverse faiths in establishing the Connecticut Hospice.[22]

Finally, hospices placed a premium on physician honesty. As early as 1847, the American Medical Association had counseled its members to refrain from giving "gloomy prognostications." During the twentieth century, the widening gulf between physician and lay knowledge encouraged further secrecy. Patients and families often participated actively in concealment. Many relatives urged doctors to withhold bad news from patients; one daughter later recalled that her entire family agreed to a "charade" throughout her mother's long cancer struggle during the 1950s. Both Flannery O'Connor and Rachel Carson attempted to hide diagnoses of fatal diseases from some friends and acquaintances as well as from the public. Other patients who intuited that death approached pretended that they remained unaware.

But we have also seen patients who protested against various types of duplicity. After learning that her mother had hidden the lupus diagnosis, O'Connor insisted on getting all medical news "straight." Because Carson's surgeon refused to inform her that her breast tumor was malignant and had metastasized, she found a physician who agreed to communicate "frankly." Although respectful of the desire of some individuals not to learn the full details of their conditions, hospice leaders noted that the dying could prepare for death only if they understood its imminence. Accurate information also was a prerequisite for participating in medical decisions. Even people closest to death, it was asserted, could retain a sense of personal mastery by exercising final authority over their own care.[23]

The beginning of the hospice movement coincided not only with various social protest movements, but also with the increased use of aggressive medical care at the end of life. During the late 1960s and early 1970s, life-sustaining technologies began to fill intensive care units.[24] As fears about the excesses of those technologies accelerated, a variety of reform efforts began. The 1990 Patient Self-Determination Act requires federally funded medical institutions (virtually all hospitals and nursing homes) to provide information about advance directives, including do-not-resuscitate orders, living wills, and health care proxies and durable powers of attorney. A few courts have issued decisions permitting the withdrawal of life supports from patients in "persistent vegetative" states. After the passage of Oregon's 2008 Death with Dignity Act, two other states have legalized physician-assisted suicide for terminally ill patients meeting certain criteria.

Those measures have proved largely ineffective. In 1998 the Robert Wood Johnson Foundation inaugurated the massive Study to Understand Prognosis and Preferences for Outcomes and Risks of Treatment (commonly known as SUPPORT), designed to improve hospital care for the dying. The report found that dying people continued to experience high levels of pain, that patient preferences for the location of death were often disregarded, and that advance directives failed to alter physicians' behavior.[25] The results of a year-long study by the Institute of Medicine, *Approaching Death: Improving Care at the End of Life*, published in 1997, concluded that "if physician and hospital performance in infection control were as poor as it is" in care for the dying, "the ensuing national outcry would create immediate demand for responses from clinicians, managers, and educators."[26] Among the "very serious problems" the report identified were inadequate pain control and a lack of physician understanding of the needs and desires of dying people.[27] More recently, researchers at Dartmouth Medical School released the Dartmouth Atlas of Health Care, demonstrating that the utilization level of health care resources in different geographic regions depends primarily on the supply of physician and hospital services. Terminally ill patients who use more resources have a shorter survival time than others, and their families report less satisfaction.[28]

"If more high-technology medicine applied to care of the seriously ill neither improves care, prolongs life, nor increases satisfaction of patients, families, and their physicians," asks Dr. Diane E. Meier of New York's Mount Sinai Medical Center, "why are we spending so much money on it?"[29] Meier points in part to the technology imperative; if the technology exists, it must be used.[30] Another common response indicts insurance incentives that reward physicians for administering procedures but not for talking with patients.[31]

A more compelling answer lies in the continuation of patterns that predated the advent of sophisticated life-sustaining technologies and the spread of insurance coverage for physician services. We have repeatedly seen patients and families refuse to relinquish hope long after they might have been expected to do so. Relatives of terminally ill persons discharged from New York's Presbyterian Hospital rejected any suggestion that these patients enter homes for incurables because that would mean giving up the struggle to live. Despite Rachel Carson's ambivalence about postwar medical science and her understanding that she, like the monarchs, was on a "closing journey," she flew to Cleveland for one last-ditch treatment shortly before her death. It is thus perhaps unsurprising that many patients today refuse to consider hospice care until too late to receive its benefits.[32] David Rieff notes that his mother, Susan Sontag, another cancer patient, was "crippled by the fear of extinction."[33] As a result, she engaged in a desperate attempt to stave off death, even traveling across the country for a bone marrow transplant that had only a remote chance of success.

According to Rieff, Sontag could never "have 'heard" that she was dying."[34] In other cases, physician evasiveness encourages late fights for survival. Doctors today are far more likely than their predecessors to reveal grim diagnoses, but most continue to withhold poor prognoses. "We are particularly uncomfortable with sharing the news that a cure is unlikely," acknowledges one physician-researcher.[35] Doctors' explanations for concealment have a familiar ring. They do not want to extinguish hope, have received little training in how to deliver bad news, and view the disclosure of a poor prognosis as an admission of failure.[36]

More fundamentally, I argue that especially after the transformation of health care at the turn of the twentieth century, medicine's imperative to cure disease consistently trumped the imperative to care for patients whose lives slowly waned. Preserving life remains the primary goal. "If you come to this hospital, we're not going to let you die," promised Dr. David T. Feinberg in 2009. Feinberg was CEO of the Ronald Reagan UCLA Medical Center, a facility distinguished by its high-intensity approach to medicine and the high amount of money it spends on patients in the last year of life.[37] Just as Ernst Boas condemned early-twentieth-century homes for incurables for admitting people whom medical science still could save, so Feinberg suggests that hospitals that devote fewer resources to gravely ill patients consign them to death prematurely. And, indeed, as nurses in the first hospital intensive care units discovered, it is often exceedingly difficult to determine which critically ill patients can be "salvaged." Some researchers highlight the incalculable benefits high-tech procedures at the end of life bestow on patients who do survive.[38] Others, however, focus on the futility and expense of those procedures as well as the additional suffering they inflict.[39]

The overriding emphasis on saving lives also explains the low priority accorded long-term care. Dying, today, often is an extremely protracted process not only because new technologies can extend the final days or weeks but even more because many people live for years with fatal chronic conditions. As physician and policy analyst Joanne Lynn remarks, commentators continue to minimize the connection between chronic disorders and mortality: "Major publications on the prevalence and impact of chronic illness simply do not mention that the end of chronic illness is death or that the impact of chronic illness tends to become more pervasive and severe toward the end of life."[40] Nevertheless, the sufferers must be counted among the population confronting mortality. The inadequacies of the care they receive most starkly demonstrate the persistence of the past in the present. Much has changed since New York City sent poor, chronically ill patients across the East River to die in miserable hovels on Blackwell's Island. Both the open boats used to transport the dying and the buildings in which they resided have gone. After the 1935 Social Security Act, almshouses gradually disappeared throughout the country. As social workers at New York's Presbyterian Hospital frequently reported during the postwar period, however, many nursing homes suffered from the problems that traditionally had plagued almshouses. Nearly twenty-five years after the Nursing Home Reform Act (or OBRA 1987) attempted to improve standards of care, large numbers of residents continue to receive substandard treatment.[41]

Moreover, despite the passage of Medicare and Medicaid, some groups continue to bear the costs of long-term care. The annual fee for a nursing home stay in 2009 was $70,000, exceeding the incomes of most people with chronic conditions.[42] Because Medicare is based on an acute-care model, it pays nursing home bills only under very restricted conditions and for no more than 100 days. Although Medicaid funds more than half of total nursing home revenues, residents must first exhaust their financial resources in order to qualify. Some evidence also suggests that many nursing homes discriminate against Medicaid patients. Because the Medicaid reimbursement rate is lower than the amount nursing homes charge private-pay residents, facilities prefer clients who can afford to pay out of pocket. Medicaid beneficiaries who do gain admission to nursing homes tend to be relegated to institutions that, according to some measures, offer the poorest quality of care.[43]

Public funding for community- and home-based services has grown rapidly in recent years but remains relatively meager. Medicare continues to emphasize medically oriented care, not the social support services many chronic care patients need. Far more of Medicaid's budget goes to institutional services than to noninstitutional ones. As a result, families provide nearly three-fourths of all long-term care.[44] Studies consistently show that although caregiving can be a profound human experience, it often causes serious physical, emotional, social, and financial problems.[45] Many people without available kin continue to struggle alone without help of any kind.

Since the 1920s, statisticians, physicians, and public health officers repeatedly have cautioned that the control of acute illnesses results in a dramatic increase in the proportion of deaths from chronic conditions and that the health care system therefore must devote far more resources to those sufferers. As baby boomers reach the age when chronic diseases are most likely to accumulate, it becomes especially urgent to heed that warning.

Abbreviations

AMLC	American Memory, Library of Congress, Washington, DC
CSS	Community Service Society Records, Rare Book and Manuscript Library, Columbia University, New York
DSD	Dorothy Smith Dushkin Papers, 1906–1989, Sophia Smith Collection, Smith College Libraries, Northampton, Massachusetts
HL	Henry E. Huntington Library, San Marino, California
JMH	John Moffat Howe Diary (1837–38), vol. 4 (New York Hospital), New-York Historical Society, New York
LACBS	Files of the Los Angeles County Board of Supervisors, Hall of Administration, Los Angeles, California
ML	Mary Lasker Papers, Rare Book and Manuscript Library, Columbia University, New York
MP	Mayors' Papers, Municipal Archives, New York
PF	Parker Family Letters, in the possession of Marianne Parker Brown, Santa Monica, California
PHPR	Presbyterian Hospital Patient Records, Archives and Special Collections, A. C. Long Health Sciences Library, Columbia University, New York
RHL	Rose Hawthorne Lathrop Papers, Dominican Sisters of Hawthorne Archives, Hawthorne, New York
SMN	Sigmund and Margaret Nestor Papers, 1942–45, New-York Historical Society, New York
TTT	Thomas Thompson Trust Records, 1844–2001, Sophia Smith Collection, Smith College Libraries, Northampton, Massachusetts

Introduction

1. "Katharine Sturgis—Pacesetter in Preventive Medicine," in *In Her Own Words: Oral Histories of Women Physicians*, ed. Regina Markell Morantz, Cynthia Stodola Pomerleau, and Carol Hansen Fenichel (New Haven: Yale University Press, 1982), p. 61.

2. See Kenneth M. Ludmerer, *Time to Heal: American Medical Education from the Turn of the Century to the Era of Managed Care* (New York: Oxford University Press, 1999), p. 134; "World Bank, World Development Indicators," Nov. 1, 2011, www.google.com/public data/explore?ds=d5bncppjof8f9.

3. Cited in Robert Jay Lifton, *Super Power Syndrome: America's Apocalyptic Confrontation with the World* (New York: Thunder's Mouth Press / Nation Books, 2003), p. 127.

4. See especially John Harley Warner and James M. Edmonson, *Dissection: Photographs of a Rite of Passage in American Medicine, 1880–1930* (New York: Blast Books, 2009); Michael Sappol, *A Traffic in Dead Bodies: Anatomy and Embodied Social Identity in Nineteenth-Century America* (Princeton: Princeton University Press, 2002).

5. See Robert A. Burt, *Death Is That Man Taking Names: Intersections of American Medicine, Law, and Culture* (Berkeley: University of California Press, 2002).

6. Philippe Ariès, *The Hour of Our Death*, trans. Helen Weaver (New York: Vintage Books, 1981), pp. 570–71.

7. Ernst Troeltsch, *Protestantism and Progress*, trans. W. Montgomery (New York: G. P. Putnam's, 1912); Max Weber, *The Protestant Ethic and the Spirit of Capitalism*, trans. Talcott Parsons (New York: Scribner, 1958).

8. See, e.g., David Gary Shaw, "Modernity between Us and Them: The Place of Religion within History," *History and Theory*, Theme Issue, 45 (Dec. 2006): 1–9; Darren E. Sherkat and Christopher G. Ellison, "Recent Developments and Current Controversies in the Sociology of Religion," *American Review of Sociology* 25 (1999): 363–94; William H. Swatos, Jr., and Kevin J. Christiano, "Secularization Theory: The Course of a Concept," *Sociology of Religion* 60, no. 3 (Autumn 1999): 209–28; R. Stephen Warner, "Work in Progress toward a New Paradigm for the Sociological Study of Religion in the United States," *American Journal of Sociology* 98, no. 5 (Mar. 1992): 1044–93.

9. Cited in Swatos and Christiano, "Secularization Theory," pp. 215–16.

10. See, e.g., Alfred F. Connors, Jr., et al., "A Controlled Trial to Improve Care for Seriously Ill Hospitalized Patients: The Study to Understand Prognoses and Preferences for Outcomes and Risks of Treatments (SUPPORT)," *Journal of the American Medical Association (JAMA)* 274, no. 20 (1995): 1591–98; Sharon R. Kaufman, *. . . And a Time to Die: How American Hospitals Shape the End of Life* (New York: Scribner, 2005); David Wendell Moller, *On Death without Dignity: The Human Impact of Technological Dying* (New York: Baywood, 1990). See also Stefan Timmermans, *Sudden Death and the Myth of CPR* (Philadelphia: Temple University Press, 1999), pp. 25–27.

11. Michael Bliss, *William Osler: A Life in Medicine* (New York: Oxford University Press, 1999), p. 291.

12. Bliss, *William Osler*; see Paul S. Mueller, "William Osler's Study of the Act of Dying: An Analysis of the Original Data," *Journal of Medical Biography* 15, Suppl. 1 (2007): 59–60. See also Richard L. Golden, "Sir William Osler" Humanistic Thanatologist," *Omega* 36, no. 3 (1997–98): 241–58.

13. Bliss, *William Osler*, p. 292.

14. See Jason Szabo, *Incurable and Intolerable: Chronic Disease and Slow Death in Nineteenth-Century France* (New Brunswick, NJ: Rutgers University Press, 2009).

15. Bliss, *William Osler*, p. 323.

16. Ibid., p. 328.

17. Daniel M. Fox, *Power and Illness: The Failure and Future of American Health Policy* (Berkeley: University of California Press, 1993); death percentages on p. 33.

18. Commission on Chronic Illness, *Chronic Illness in the United States*, vol. 2: *Care of the Long-Term Patient* (Cambridge: Harvard University Press, 1956), p. 21.

19. Morton L. Levin, "A Call for Action," in Leonard W. Mayo et al., *Chronic Illness: National Health Forum*, Public Health Reports 71, no. 7 (July 1956): 695.

20. "Planning for the Chronically Ill," *American Journal of Public Health* 37, no. 10 (Oct. 1947): 1264.

21. Leonard W. Mayo, "Five Million People," in *Chronic Illness: National Health Forum*, p. 678.

22. Commission on Chronic Illness, *Chronic Illness in the United States*, p. 133.

23. Edward D. Berkowitz, *Disabled Policy: America's Programs for the Handicapped* (New York: Cambridge University Press, 1987), pp. 155–83; Ruth O'Brien, *Crippled Justice: The History of Modern Disability Policy in the Workplace* (Chicago: University of Chicago Press, 2001).

24. *The Human Radiation Experiments: Final Report of the President's Advisory Committee* (New York: Oxford University Press, 1996), p. 179; *Terminal Care for Cancer Patients: A Survey of the Facilities and Services Available and Needed for the Terminal Care of Cancer Patients in the Chicago Area* (Chicago: Central Service for the Chronically Ill of the Institute of Medicine of Chicago, 1950).

25. Mary Ann Krisman-Scott, "An Historical Analysis of Disclosure of Terminal Status," *Journal of Nursing Scholarship* 32, no. 1 (Mar. 2000): 47–52.

26. See, e.g., Diane E. Meier, "The Development, Status, and Future of Palliative Care," www.rwjf.org/files/research/4588.pdf, accessed Nov. 10, 2011.

27. See Elisabeth Kübler-Ross, *On Death and Dying* (New York: Scribner, 1969). For more recent examples, see Atul Gawande, "Letting Go," *New Yorker*, Aug. 2, 2010; Donald Joralemon, "Mortal Dilemmas: Why Is It So Bad to Die in America?" http://sophia.smith.edu/blog/mortaldilemmas, accessed Nov. 3, 2011.

Chapter 1 · The Good Death at Home

1. Mary Ann Webber to Very Dear Children, May 1, 1867, PF.

2. Mary Ann Webber, early spring, 1867, PF.

3. Mary Ann Webber, Wednesday morn, n.d., PF.

4. Mary Ann Webber to Dear Ones, Oct. 28, 1861, PF.

5. Mary Ann Webber to Mary, sometime in 1862, PF.

6. *Frontier Mother: The Letters of Gro Svendsen*, ed. Pauline Farseth and Theodore C. Blegen (Northfield, MN: Norwegian-American Historical Association, 1950), p. 136.

7. *Mrs. Longfellow: Selected Letters and Journals of Fanny Appleton Longfellow (1817–1861)*, ed. Edward Wagenknecht (New York: Longmans, Green, 1956), p. 142.

8. "The Conine Family Letters, 1849–1851: Employed in Honest Business and Doing the Best We Can," ed. Donald E. Baker, *Indiana Magazine of History* 69, no. 4 (1974): 177.

9. Jane R. Pomeroy, *Alexander Anderson (1775–1870), Wood Engraver and Illustrator: An Annotated Bibliography*, vol. 1 (New York: American Antiquarian Society and the New York Public Library, 2005), p. xxx. Anderson subsequently became a famous wood engraver.

10. *"A Secret to Be Burried": The Diary and Life of Emily Hawley Gillespie, 1858–1888*, ed. Judy Nolte Lensink (Iowa City: University of Iowa Press, 1989), pp. 17–18.

11. Excerpts from diary of Sarah Jane Price in Mary Hurlbut Cordier, *Schoolwomen of the Prairies and Plains: Personal Narratives from Iowa, Kansas, and Nebraska, 1860s-1920s* (Albuquerque: University of New Mexico Press, 1992), p. 190.

12. *Diary of Sarah Connell Ayer* (Portland, ME: Lefavor-Tower, 1910), pp. 244–45.

13. "A Young Woman in the Midwest: The Journal of Mary Sears, 1859–1860," *Ohio History* 82, nos. 3–4 (Summer/Autumn 1970): 228.

14. Nannie Stillwell Jackson, *Vinegar Pie and Chicken Bread: A Woman's Diary of Life in the Rural South, 1890–1891*, ed. Margaret Jones Bolsteri (Fayetteville: University of Arkansas Press, 1982), pp. 66–67, 88–100.

15. Laura I. Oblinger to Uriah W. Oblinger, Jan. 29, 1882, "Prairie Settlement: Nebraska Photographs and Family Letters, 1862–1912," Nebraska State Historical Society [Digital ID nbhips 1224].

16. *Little Verses for Good Children*, p. 9, Sunday School Books: Shaping the Values of Youth in Nineteenth-Century America, AMLC.

17. Drew Gilpin Faust, *This Republic of Suffering: Death and the American Civil War* (New York: Knopf, 2008); Shai J. Lavi, *The Modern Art of Dying: A History of Euthanasia in the United States* (Princeton: Princeton University Press, 2005); Mark S. Schantz, *Awaiting the Heavenly Country: The Civil War and America's Culture of Death* (Ithaca, NY: Cornell University Press, 2008).

18. Eliza Webber to Parents, July 22, 1863, PF.

19. Mary Ann Webber to Mary, May 24, 1871, PF.

20. Mary Ann Webber to My Dear Children, June 11, 1871, PF.

21. Mary Ann Webber to Dear Children, June 15, 1871, PF.

22. Mary Ann Webber to My Dear Children, June 15, 1871, PF.

23. Mary Ann Webber to Alpha, Aug. 16, 1871, PF.

24. Sheila M. Rothman, *Living in the Shadow of Death: Tuberculosis and the Social Experience of Illness in American History* (Baltimore: Johns Hopkins University Press, 1995), p. 13.

25. Pat Jalland, *Death in the Victorian Family* (New York: Oxford University Press, 1996), p. 41; see also Schantz, *Awaiting the Heavenly Country.*

26. Henry Guernsey to Emmeline Donaldson Guernsey, Feb. 6, 1876, and Nov. 19, 1876, file 32, box 2, Guernsey Family Papers, 1837–1957, HL.

27. Henry Guernsey to Emmeline Donaldson Guernsey, July 28, 1877, file 32, box 2, Guernsey Family Papers. Throughout, I have retained the spellings and punctuation (or lack thereof) of manuscript materials.

28. *Sam Curd's Diary: The Diary of a True Woman*, ed. Susan S. Arpad (Athens: Ohio University Press, 1984), pp. 103–26.

29. Faust, *Republic of Suffering*, p. 10.

30. Diary of Martha Shaw Farnsworth, Oct. 26, 1893, Martha Farnsworth Collection, Kansas State Historical Society, Topeka, KS.

31. Mary Adams to Eliza, Jan. 19, 1866, PF.

32. "Private Journal of Mary Ann Owen Sims," pt. 1, ed. Clifford Dale Whitman, *Arkansas Historical Quarterly* 35, no. 4 (Winter 1976): 150.

33. Ibid., p. 176.

34. *Diary of Sarah Connell Ayer*, p. 274.

35. "Young Woman in the Midwest," p. 229.

36. "Letters from the Past," *Vermont Quarterly: A Magazine of History* 20, no. 1 (Jan. 1952): 49–50.

37. Lawrence Stone, "Death in New England," *New York Review of Books* 25, no. 16 (Oct. 26, 1978).

38. Faust, *Republic of Suffering*, p. 31.

39. Mary Ann Webber, n.d., PF.

40. Mary Ann Webber, July 23, 1871, PF.

41. Mary Ann Webber, June 11, 1871, PF.

42. Mary Ann Webber to son, Aug. 16, 1871, PF.

43. *Louisa May Alcott: Her Life, Letters, and Journals*, ed. Ednah D. Cheney (Boston: Roberts Brothers, 1892), p. 300.

44. Jane Freeland to sister and family, *Read This Only to Yourself: The Private Writings of Midwestern Women, 1880–1910*, ed. Elizabeth Hampsten (Bloomington: Indiana University Press, 1982), p. 83.

45. See Philippe Ariès, *Western Attitudes toward Death: From the Middle Ages to the Present*, trans. Patricia M. Ranum (Baltimore: Johns Hopkins University Press, 1974), p. 570; Jeanne C. Quint, *The Nurse and the Dying Patient* (New York: Macmillan, 1967).

46. See Clark Lawlor, *Consumption and Literature: The Making of the Romantic Disease* (New York: Palgrave Macmillan, 2006).

47. *Matilda's Letters*, ed. Barbara Trueblood Abbott (privately printed, 1974), p. 84.

48. Eliza W. Farnham, *Life in Prairie Land* (New York: Harper, 1846), p. 245.

49. *The Diary of Ellen Birdseye Wheaton*, with notes by Donald Gordon (Boston: privately printed, 1923), p. 231.

50. Jane Freeland to sister and family, *Read This Only to Yourself*, p. 83.

51. Quoted in Hampsten, *Read This Only to Yourself*, pp. 140–41.

52. Helen Smith Jordan, *Love Lies Bleeding* (privately printed, 1979), p. 481.

53. Quoted in Michael Sappol, *A Traffic of Dead Bodies: Anatomy and Embodied Social Identity in Nineteenth-Century America* (Princeton: Princeton University Press, 2002), p. 78.

54. *Memoirs of Catharine Seely and Deborah S. Roberts, Late of Darien, Connecticut* (New York: Daniel Godwin, 1844), p. 48.

55. Farnham, *Life in Prairie Land*, p. 250.

56. Excerpts from diary of Sarah Jane Price in Cordier, *Schoolwomen of the Prairies and Plains*, p. 197.

57. Laura Oblinger to Uriah Oblinger, Jan. 29, 1882, Nebraska State Historical Society.

58. See Ann Douglas, *The Feminization of American Culture* (New York: Anchor Press / Doubleday, 1988); Karen Halttunen, *Confidence Men and Painted Women: A Study of Middle-Class Culture in America, 1830–1870* (New Haven: Yale University Press, 1982); Gary Landerman, *The Sacred Remains: American Attitudes toward Death, 1799–1883* (New Haven: Yale University Press, 1996), pp. 60–61; Wendy Simonds and Barbara Katz Rothman, *Centuries of Solace: Expressions of Maternal Grief in Popular Literature* (Philadelphia: Temple University Press, 1992).

59. Gary Scott Smith, *Heaven in the American Imagination* (New York: Oxford, 2011), p. 70.

60. Mary Ann Webber to Emma, Mar. 17, 1870, PF; see Smith, *Heaven in the American Imagination*, p. 71.

61. *Memoirs of Catharine Seely and Deborah S. Roberts*, p. 231.

62. *Diary of Sarah Connell Ayer*, p. 215.

63. *Diary of Ellen Birdseye Wheaton*, p. 231.

64. Halttunen, *Confidence Men and Painted Women*, pp. 129–130.

65. Amelia Akehurst Lines, *To Raise Myself a Little: The Diaries and Letters of Jennie, a Georgia Teacher, 1851–1886*, ed. Thomas Dyer (Athens: University of Georgia Press, 1982), p. 205.

66. "The Letters of the Rev. and Mrs. Olof Olsson, 1869–1873, Pioneer Founders of Lindsborg," ed. Emory Lindquist, *Kansas Historical Quarterly* 21, no. 7 (Autumn 1955): 511.

67. *Caleb and Mary Wilder Foote: Reminiscences and Letters*, ed. Mary Wilder Tileston (Boston: Houghton Mifflin, 1918), pp. 55–56.

68. Ibid., p. 63.

69. Quoted in Hampsten, *Read This Only to Yourself*, p. 141.

70. "Diary of Mrs. Joseph Duncan," ed. Elizabeth Duncan Putnam, *Journal of the Illinois State Historical Society* 21, no. 1 (April 1928): 75.

71. Susan I. Lesley, *Recollections of My Mother, Mrs. Anne Jean Lyman, of Northampton; Being a Picture of Domestic and Social Life in New England in the First Half of the Nineteenth Century* (Boston: Houghton, Mifflin, 1899), p. 315.

72. Mary Ann Webber to friends, Apr. 29, 1863, PF.

73. Mary Ann Webber to Mary, Apr. 30, 1863, PF.

74. Mary Ann Webber to son, Apr. 1863, PF.

75. *The Diaries of Sally and Pamela Brown, 1832–1838, Plymouth Notch, Vermont*, ed. Blanche Brown Bryant and Gertrude Elaine Baker (Springfield, VT: William L. Bryant Foundation, 1970), p. 52.

76. Lines, *To Raise Myself a Little*, p. 199.

77. Quoted in Sally G. McMillen, *Motherhood in the Old South: Pregnancy, Childbirth, and Infant Rearing* (Baton Rouge: Louisiana State University Press, 1990), p. 174.

78. *Caleb and Mary Wilder Foote*, p. 95.

79. Joan D. Hedrick, *Harriet Beecher Stowe: A Life* (New York: Oxford University Press, 1994), p. 281.

80. Quoted in Joan D. Hedrick, "'Peaceable Fruits': The Ministry of Harriet Beecher Stowe," *American Quarterly* 40, no. 3 (Sept. 1988): 307–22.

81. Quoted in Hampsten, *Read This Only to Yourself*, p. 141.

82. Lines, *To Raise Myself a Little*, p. 207.

83. *Life and Letters of Catharine M. Sedgwick*, ed. Mary E. Dewey (New York: Harper, 1871), p. 110.

84. Mary Adams to Albert Denny, Feb. 13, 1881, PF.

Chapter 2 • Medical Professionals (Sometimes) Step In

1. Henry A. Guernsey to Sarah Guernsey Beebe, and Guernsey to Emmeline Guernsey, both dated Jan. 14, 1877, file 30, box 2, Guernsey Family Papers, HL

2. Quoted in Judy Nolte Lensink, *"A Secret to Be Burried": The Diary and Life of Emily Hawley Gillespie, 1858–1888* (Iowa City: University of Iowa Press, 1989), p. 11.

3. Margaret Jones Bolsteri, *Vinegar Pie and Chicken Bread: A Woman's Diary of Life in the Rural South, 1890–1891* (Fayetteville: University of Arkansas Press, 1982), p. 99.

4. Interview with William G. Urton, Boswell, NM, American Life Histories: Manuscripts from the Federal Writers' Project, 1936–1940, AMLC.

5. Lamar Riley Murphy, *Enter the Physician: The Transformation of Domestic Medicine, 1760–1860* (Tuscaloosa: University of Alabama Press, 1911), pp. 32–69.

6. Alta Marvey Heiser, *Quaker Lady: The Story of Charity Lynch* (Oxford, OH: Mississippi Valley Press, 1941), p. 135.

7. Harriet Connor Brown, *Grandmother Brown's Hundred Years, 1827–1927* (New York: Blue Ribbon Books, 1929), pp. 152–53.

8. See, e.g., Samuel J. Crumbine, *Frontier Doctor* (Philadelphia: Dorrance, 1948), p. 62.

9. See William G. Rothstein, *American Physicians in the Nineteenth Century: From Sects to Science* (Baltimore: Johns Hopkins University Press, 1985), pp. 288–89.

10. Cited in Paul Starr, *The Social Transformation of American Medicine: The Rise of a Sovereign Profession and the Making of a Vast Industry* (New York: Basic Books, 1982), p. 118.

11. J. Marion Sims, *The Story of My Life* (1881; repr., New York: Da Capo, 1968), pp. 139–43.

12. Charles Beneulyn Johnson, *Sixty Years in Medical Harness; or, The Story of a Long Medical Life, 1865–1925* (New York: Medical Life Press, 1926), pp. 212–13.

13. William Allen Pusey, *A Doctor of the 1870's and 80's* (Springfield, IL: Charles D. Thomas, 1932), p. 94.

14. See M. Steven Stowe, "Obstetrics and the Work of Doctoring in the Mid-Nineteenth-Century American South," *Bulletin of the History of Medicine* 64, no. 2 (Fall 1990): 543.

15. Quoted in Steven M. Stowe, *A Southern Practice: The Diary and Autobiography of Charles A. Hentz, M.D.* (Charlottesville: University Press of Virginia, 2000), p. 325.

16. Ibid., pp. 334, 336.

17. See Ian Mortimer, *The Dying and the Doctors: The Medical Revolution in Seventeenth-Century England* (Rochester, NY: Boydell Press, 2009); Roy Porter, "Death and the Doctors in Georgian England," in *Death, Ritual, and Bereavement*, ed. Ralph Houlbrooke (New York: Routledge, 1989), pp. 77–94.

18. American Medical Association, *Medical Ethics and Etiquette: The Code of Ethics Adopted by the American Medical Association*, with commentaries by Austin Flint (New York: Appleton, 1895), p. 24.

19. Ibid.

20. W. Bruce Fye, "Active Euthanasia: An Historical Survey of Its Conceptual Origins and Introduction into Medical Thought," *Bulletin of the History of Medicine* 52, no. 4 (1979): 496; Shai J. Lavi, *The Modern Art of Dying: A History of Euthanasia in the United States* (Princeton: Princeton University Press, 2005), p. 61; Roselyne Rey, *A History of Pain*, trans. Louise Elliott Wallace, J. A. Cadden, and S. W. Cadden (Cambridge: Harvard University Press, 1995).

21. Mary Richardson Walker, "Diary of Mary Richardson Walker, November 1877," in *Mary Richardson Walker: Her Book* (Caldwell, ID: Caxton Printers, 1945), p. 357.

22. Louisa May Alcott to Mrs. A. D. Mosher, Dec. 16, 1877, in *The Selected Letters of Louisa May Alcott*, ed. Joel Myerson and Daniel Shealy (Boston: Little, Brown, 1987), p. 227.

23. Lavi, *Modern Art of Dying*, p. 61.

24. John Matteson, *Eden's Outcasts: The Story of Louisa May Alcott and Her Father* (New York: Norton, 2007), p. 235.

25. Louisa May Alcott to Eliza Wells, Mar. 19, 1858, in Myerson and Shealy, *Selected Letters*, p. 33.

26. "A Young Woman in the Midwest: The Journal of Mary Sears, 1859–1860," *Ohio History* 82, nos. 3 & 4 (Summer–Autumn 1973): 228.

27. AMA, *Medical Ethics and Etiquette*, p. 21.

28. Ibid., p. 23.

29. Worthington Hooker, *Physician and Patient; or, A Practical View of the Mutual Duties, Relations, and Interests of the Medical Profession and the Community* (1849; repr., New York: Arno Press, 1972), pp. 354–55.

30. Quoted in Charles E. Rosenberg, "Therapeutic Revolution: Medicine, Meaning, and Social Change in Nineteenth-Century America," in *Sickness and Health in America: Readings in the History of Medicine and Public Health*, ed. Judith Walzer Leavitt and Ronald L. Numbers (Madison: University of Wisconsin Press, 1985), p. 48.

31. AMA, *Medical Ethics and Etiquette*, p. 18.

32. Nicholas A. Christakis, *Death Foretold: Prophecy and Prognosis in Medical Care* (Chicago: University of Chicago Press, 1999), p. xix.

33. Caroline Clapp Briggs, *Reminiscences and Letters of Caroline C. Briggs*, ed. George S. Merriam (Boston: Houghton, Mifflin, 1897), p. 445.

34. Martha Saxton, *Louisa May: A Modern Biography of Louisa May Alcott* (New York: Avon Books, 1978).

35. Mary Serena Eliza Blair to Violet Blair Janin, Mar. 11, 1875, file 1, box 4, Janin Family Letters, HL.

36. *The Diary of Emily Jane Green Hollister: Her Nursing Experiences*, ed. Deborah D. Smith (Ann Arbor: Historical Center for the Health Sciences, University of Michigan, 1991), pp. 43, 49, 65.

37. Ibid., pp. 37–38, 43.

38. Jan. 14, 1838, JMH.

39. Oct. 5, 1837, JMH.

40. Nov. 19, 1837, JMH.

41. Apr. 15, 1838, JMH.

42. Apr. 17, 1838, JMH.

43. Apr. 22, 1838, JMH.

44. St. Luke's Hospital, *Eighth Annual Report for the Year Ending St. Luke's Day, Oct. 18, 1866* (St. Johnland, NY: Orphan Boys' Press, 1867), pp. 8–9. For a Philadelphia example, see Charles E. Rosenberg, *Care of Strangers: The Rise of America's Hospital System* (New York: Basic Books, 1987), p. 113.

45. See Alan M. Kraut, *Silent Travelers: Germs, Genes, and the "Immigrant Menace"* (New York: Basic Books, 1994), pp. 44–46; Bernadette McCauley, *Who Shall Take Care of Our Sick? Roman Catholic Sisters and the Development of Catholic Hospitals in New York City* (Baltimore: Johns Hopkins University Press, 2005); Sioban Nelson, *Say Little, Do Much: Nursing, Nuns, and Hospitals in the Nineteenth Century* (Philadelphia: University of Pennsylvania Press, 2001); Barbra Mann Wall, *Unlikely Entrepreneurs: Catholic Sisters and the Hospital Marketplace, 1865–1925* (Columbus: Ohio State University Press, 2005).

46. Barbra Mann Wall and Sioban Nelson, "Our Heels are Praying Very Hard All Day," *Holistic Nursing Practice*, Nov./Dec. 2003, pp. 320–28.

47. Rosenberg, *Care of Strangers*, p. 18.

48. Morris J. Vogel, *The Invention of the Modern Hospital* (Chicago: University of Chicago Press, 1980), p. 1.

49. Rosenberg, *Care of Strangers.*

50. David Rosner, *A Once Charitable Enterprise: Hospitals and Health Care in Brooklyn and New York, 1885–1915* (Princeton: Princeton University Press, 1982), p. 17; Starr, *The Social Transformation of American Medicine.*

51. Rosenberg, *Care of Strangers*, pp. 56, 116.

52. Ibid., pp. 23, 116; Vogel, *Invention of the Modern Hospital*, pp. 8, 335–36.

53. Amos G. Warner, *American Charities: A Study in Philanthropy and Economics* (New York: Crowell, 1894), p. 242.

54. Thomas G. Morton with Frank Woodbury, *History of Pennsylvania Hospital, 1751–1895* (Philadelphia: Times Printing House, 1895), p. 210.

55. Susan Sontag, *Illness as Metaphor* (New York: Vintage Books, 1979), pp. 11–12.

56. Quoted in Deborah Kuhn McGregor, *From Midwives to Medicine: The Birth of American Gynecology* (New Brunswick, NJ: Rutgers University Press, 1998), p. 192.

57. Morton and Woodbury, *History of the Pennsylvania Hospital*, p. 210.

58. Brooklyn Homeopathic Hospital, *By-Laws and Regulations of the Brooklyn Homeopathic Hospital, Adopted by the Board of Trustees, 1871* (New York: George F. Nesbitt, 1871), p. 12. Cf. Hartford Hospital, *Fourteenth Annual Report of the Executive Committee of the Hartford Hospital for the Year 1874* (Hartford, CT: Hartford Hospital, 1874), pp. 10–11.

59. McGregor, *From Midwives to Medicine*, p. 197; Regina Morantz-Sanchez, *Conduct Unbecoming a Woman: Medicine on Trial in Turn-of-the-Century Brooklyn* (New York: Oxford University Press, 1999), p. 12.

60. For an English example of hospitals ignoring their own policies with regard to tuberculosis patients, see Graham Mooney, Bill Luckin, and Andrea Tanner, "Patient Pathways: Solving the Problem of Institutional Mortality in London during the Later Nineteenth Century," *Social History of Medicine* 12, no. 2 (1999): 227–69.

61. John Allan Hornsby and Richard E. Schmidt, *The Modern Hospital: Its Inspiration, Its Architecture, Its Equipment, Its Operation* (Philadelphia: W. B. Saunders, 1914), p. 120; Starr, *Social Transformation*, p. 157.

62. Rosenberg, *Care of Strangers*, p. 31.

63. Quoted in Vogel, *Invention of the Modern Hospital*, p. 4.

64. See, e.g., Brooklyn Homeopathic Hospital, *By-Laws and Regulations, 1871*, p. 22; New Britain General Hospital, New Haven, Connecticut, *First Annual Report, for the Year Ending May 1, 1900* (New Haven: New Britain General Hospital, 1900), p. 43; Massachusetts General Hospital, *Acts, Resolves, By-Laws, and Rules and Regulations* (Boston: James Loring, 1837), p. 32.

65. Minutes of the Executive Committee, Montefiore Home, A Hospital for Chronic Invalids, May 15, 1892, Harry M. Zimmerman Archives, Montefiore Medical Center, Bronx, New York.

66. Apr. 17, 1838, JMH.

67. E. R. Peaslee, R. A. Emmet, and T. G. Thomas, *Reply to Dr. J. Marion Sims' Pamphlet Entitled "The Woman's Hospital in 1874"* (New York: Trow's Printing and Bookbinding, 1877), p. 18.

68. Vogel, *Invention of the Modern Hospital*, pp. 34–36.

69. "Bellevue Hospital and the 'Transfer Cases,'" *New York Times*, Jan. 22, 1855.

70. Commissioners of Public Charities and Correction, New York, *Eighteenth Annual Report for the Year 1877* (New York: Department Press, 1878), p. 31.

71. "The Charity Hospital," *New York Times*, July 28, 1883.

72. "Blackwell's Island Outrage," *New York Times*, May 7, 1867.

73. Blackwell's Island Hospitals, New York City, *The Report of the Resident Physician of Blackwell's Island, for the Year 1857* (New York: Chas W. Baker, 1858), p. 24.

74. Michael B. Katz, *In the Shadow of the Poorhouse: A Social History of Welfare in America* (New York: Basic Books, 1986), pp. 25–29.

75. Commissioners of Public Charities and Correction, *Eighteenth Annual Report, 1877*, p. 55.

76. Homer Folks, "Reform and Public Charities," folder "Misc. Papers, 1897–1898," box 3, Homer Folks Papers, Rare Book and Manuscript Library, Columbia University, New York.

77. Commissioners of Public Charities and Correction, New York, *Twelfth Annual Report for the Year 1871* (New York: Bellevue Press, 1871), p. 2.

78. Folks, "Reform and Public Charities."

79. Minutes of the Commissioner of Public Charities and Correction, Apr. 15, 1869, p. 21, in Library, New York Academy of Medicine, New York.

80. Commissioners of Public Charities and Correction, New York, *Twenty-Seventh Annual Report for the Year 1886* (New York: NYC Asylum for the Insane, Ward's Island, 1887), p. 22.

Chapter 3 · Cultivating Detachment, Sidetracking Care

Portions of this chapter appeared in an earlier form as " 'In the Last Stages of Irremediable Disease': American Hospitals and Dying Patients before World War II," *Bulletin of the History of Medicine* 85, no. 1 (Spring 2011): 29–56. Used with the generous permission of the *Bulletin of the History of Medicine*.

1. Rosemary Stevens, *In Sickness and in Wealth: American Hospitals in the Twentieth Century* (New York: Basic Books, 1989), p. 35.

2. Cited in Charles C. Rosenberg, *The Care of Strangers: The Rise of America's Hospital System* (New York: Basic Books, 1987), p. 5.

3. See Ronald L. Numbers, "The Rise and Fall of the American Medical Profession," in *Sickness and Health in America: Readings in the History of Medicine and Public Health*, ed. Judith Walzer Leavitt and Ronald L. Numbers (Madison: University of Wisconsin Press, 1985); Paul Starr, *The Social Transformation of American Medicine: The Rise of a Sovereign Profession and the Making of a Vast Industry* (New York: Basic Books, 1982).

4. Ellen D. Baer, "Nurses," in *Women, Health, and Medicine in America: A Historical Handbook*, ed. Rima D. Apple (New Brunswick, NJ: Rutgers University Press, 1992), p. 454. See Barbara Melosh, *"The Physician's Hand": Work Culture and Conflict in American Nursing* (Philadelphia: Temple University Press, 1982); Susan M. Reverby, *Ordered to Care: The Dilemma of American Nursing, 1850–1945* (New York: Cambridge University Press, 1987).

5. Melosh, *The Physician's Hand*, pp. 77, 92.

6. Alfred Worcester, "The Care of the Dying," in *Physician and Patient: Personal Care*, ed. L. Eugene Emerson (Cambridge: Harvard University Press, 1929), pp. 200–224.

7. Walter C. Alvarez, "Care of the Dying," *JAMA* 150 (Sept. 13, 1952): 86–91; Frank J. Ayd, Jr., "The Hopeless Case," *JAMA* 181 (Sept. 29, 1962): 1099–1102. See Christopher

Crenner, *Private Practice in the Early-Twentieth-Century Medical Office of Dr. Richard Cabot* (Baltimore: Johns Hopkins University Press, 2005).

8. See Jeanne E. Quint, *The Nurse and the Dying Patient* (New York: Macmillan, 1967).

9. Quoted in Berniece M. Wagner, "Teaching Students to Work with the Dying," *American Journal of Nursing* 64, no. 11 (Nov. 1964): 128.

10. See Rosenberg, *Care of Strangers.*

11. Kenneth M. Ludmerer, *Time to Heal: American Medical Education from the Turn of the Century to the Era of Managed Care* (New York: Oxford University Press, 1999), p. 109.

12. Condict W. Cutler, Jr., "Forty Years Ago," in *The Roosevelt Hospital, 1871–1957* (New York: Roosevelt Hospital, 1957), p. 170.

13. Michael Bliss, *The Making of Modern Medicine: Turning Points in the Treatment of Disease* (Chicago: University of Chicago Press, 2011), p. 35.

14. Robert A. Aronowitz, *Unnatural History: Breast Cancer and American Society* (New York: Cambridge University Press, 2007), p. 132.

15. Keith Wailoo, *How Cancer Crossed the Color Line* (New York: Oxford University Press, 2011), p. 31. Prescott may have felt it especially important to append that note because Cabot had a reputation for disclosing bad news to patients. See Crenner, *Private Practice*, pp. 110–13.

16. See Aronowitz, *Unnatural History*; Wailoo, *How Cancer Crossed the Color Line.*

17. Barbara Bates, *Bargaining for Life: A Social History of Tuberculosis, 1876–1938* (Philadelphia: University of Pennsylvania Press, 1992).

18. Lawrence F. Flick to H. W. F., Pennsylvania, Aug. 22, 1906, Papers of Lawrence F. Flick, Archives of the Catholic University of America, Washington, DC.

19. Emily K. Abel, *Hearts of Wisdom: American Women Caring for Kin* (Cambridge: Harvard University Press, 2000), p. 140.

20. J. V. DePorte, "Where Do People Die—at Home or in Hospitals?" *Modern Hospital* 33, no. 2 (Aug. 1929): 73.

21. Shigeaki Hinohara, "Sir William Osler's Philosophy on Death," *Annals of Internal Medicine* 118 (1993): 638–42.

22. Melosh, *The Physician's Hand*, p. 56.

23. Quoted in Bates, *Bargaining for Life*, pp. 227–28.

24. Charles Dwight Willard to Sarah Hiestand, Sept. 5, 1908, Charles Dwight Willard Collection, HL.

25. Charles Dwight Willard to May Willard, Aug. 1908, and Oct. 29, 1908, Willard Collection.

26. "Registered Nurse," South Carolina Writers' Project, Charleston, SC, Life History, Jan. 20, 1939, AMHC.

27. Karen Buhler-Wilkerson, *No Place Like Home: A History of Nursing and Home Care in the United States* (Baltimore: Johns Hopkins University Press), 2001.

28. Ada Beazley to Lillian Wald, Dec. 24, 1916, box 44, Lillian Wald Papers, Rare Book and Manuscript Library, Butler Library, Columbia University, New York.

29. Quoted in J. H. Jennett, "Why Did the Patient Die?" *Modern Hospital* 47, no. 4 (Oct. 1936): 81.

30. See, e.g., E. M. Bluestone, "The Value of Post-Mortem Examinations and Methods of Obtaining Them," *Modern Hospital* 18, no. 5 (May 1922): 413–27; Ludvig Hektoen, "Necropsy Percentage in Relation to Hospital Professional Efficiency," *Modern Hospital* 22,

no. 5 (May 1924): 491–93; Jennett, "Why Did the Patient Die?" pp. 81–83; "The Necropsy Percentage and Hospital Efficiency" (Editorial), *Modern Hospital* 22, no. 4 (Apr. 1924): 374–75; Milton Plotz, "The Jewish Attitude toward Autopsies," *Modern Hospital* 45, no. 5 (Nov. 1935): 67–68; Milton Plotz, "The Moral, Physical and Legal Aspects of Autopsies," *Modern Hospital* 37, no. 7 (July 1931): 83–87; Cyrus C. Sturgis, "Methods of Obtaining More Necropsies," *Modern Hospital* 16, no. 6 (June 1921): 497–99; John W. Williams, "Autopsy—The Answer to the Demand for Fewer Unfounded Diagnoses," *Modern Hospital* 37, no. 4 (Apr. 1932): 75–76. See also Dorothy Levenson, *Montefiore: The Hospital as Social Instrument* (New York: Farrar, Straus & Giroux, 1984); Hampton P. Howell, "Roosevelt Hospital in the Mid-Nineties," in *Roosevelt Hospital*, p. 145; John Allan Hornsby and Richard E. Schmidt, *The Modern Hospital: Its Inspiration, Architecture, Equipment, Operation* (Philadelphia: W. B. Saunders, 1913) , p. 507.

31. *Annual Report of the Carney Hospital for the Year 1895* (Boston: John Cashman, 1896), p. 10; *Thirty-sixth Annual Report of St. Francis' Hospital, Sisters of the Poor of St. Francis, for the Year Ending December 31, 1901* (West Chester, NY.: Press of New York Catholic Protectory, 1902), p. 5; Alan M. Kraut and Deborah A. Kraut, *Covenant of Care: Newark Beth Israel and the Jewish Hospital in America* (New Brunswick, NJ: Rutgers University Press, 2007).

32. Edward Fletcher Stevens, *The American Hospital of the Twentieth Century: A Treatise on the Development of Medical Institutions, both in Europe and in America, since the Beginning of the Present Century* (New York: Architectural Record Publishing, 1918), p. 1.

33. S. S. Goldwater, "On Humanizing the Hospital," *Modern Hospital* 22, no. 6 (June 1924): 540.

34. Harold J. Seymour, "What Does the Public Think about Your Hospital?" *Modern Hospital* 17, no. 6 (Dec. 1921): 481.

35. Frank W. Hoover, "The Small Hospital Executive as a Publicity Agent," *Modern Hospital* 38, no. 4 (April 1932): 81.

36. Annmarie Adams, *Medicine by Design: The Architect and the Modern Hospital, 1893–1943* (Minneapolis: University of Minnesota Press, 2008), p. 39.

37. See Janet Peterkin, "Innovations That Feature a New Hospital Building," *Modern Hospital* 37, no. 5 (May 1932): 91; Hornsby and Schmidt, *The Modern Hospital*, p. 387; Richard Resler, "Small Hospital Morgue and Autopsy Room," *Modern Hospital* 20, no. 3 (Mar. 1923): 228.

38. Stevens, *American Hospital*, p. 201.

39. Herbert J. Kellaway, "Landscape Architecture as Applied to Hospital Grounds," *Modern Hospital* 18, no. 4 (Apr. 1922): 306–7.

40. William Seaman Bainbridge, *The Cancer Problem* (New York: Macmillan, 1914), p. 433.

41. "Founded by James Lenox / The Chief Features of the Presbyterian Hospital," *New York Times*, July 3, 1892.

42. Lewis Thomas, "Dying as Failure," *Annals of the American Academy of Political and Social Science* 447 (Jan. 1980): 2.

43. "St. Luke's Inquiry into Neglect Charge," *New York Times*, Feb. 16, 1911.

44. Carlos Bulosan, *America Is in the Heart: A Personal History* (Seattle: University of Washington Press, 1943), p. 236.

45. L. J. Frank, "The Individual Room Hospital—The Hospital of the Future," *Modern Hospital* 17, no. 6 (Dec. 1921): 476.

46. "Medical Director's Report," Chestnut Hill Hospital, Philadelphia, *Twenty-Sixth Annual Report for the Year Ending December 31st 1929* (Philadelphia: Chestnut Hill Hospital, 1930), p. 8.

47. General Memorial Hospital, *Twenty-first Annual Report for the Year 1905* (New York: Knickerbocker Press, 1905), p. 16; see also Morris J. Vogel, *The Invention of the Modern Hospital: Boston, 1870–1943* (Chicago: University of Chicago Press, 1980), p. 76.

48. See Sarah Gordon, ed., *All Our Lives: A Centennial History of Michael Reese Hospital and Medical Center, 1881–1981* (Chicago: Michael Reese Hospital and Medical Center, 1981), p. 88; David Rosner, *A Once Charitable Enterprise: Hospitals and Health Care in Brooklyn and New York, 1885–1915* (New York: Cambridge University Press, 1982), pp. 19–20.

49. Babies' Hospital, New York, *Twelfth Annual Report, from Oct. 1st, 1899 to September 30th, 1900* (New York: Babies' Hospital), p. 25.

50. For a fuller account of the impact of ambulances, see Emily Abel, "Patient Dumping in New York City, 1877–1917," *American Journal of Public Health* 101, no. 5 (May 2011): 789–95.

51. Ryan Corbett Bell, *The Ambulance: A History* (Jefferson, NC: McFarland, 2009), pp. 58–79; Norman Maul, "Systematic Ambulance Service for Metropolitan Hospitals," *Modern Hospital* 4, no. 1 (Jan. 1915): 31–32.

52. Quoted in Bell, *The Ambulance*, p. 67.

53. "Little Julia Bictor's Death," *New York Times*, Dec. 7, 1884.

54. "Death Caused by Removal," *New York Times*, Aug. 10, 1891.

55. "Long Wait for Ambulance," *New York Times*, Jan. 28, 1897.

56. *Annual Report of the Department of Public Charities of the City of New York, 1902* (New York: Mail and Express Co., 1903), p. 24.

57. "To End Transfers of Dying," *New York Times*, Mar. 7, 1906.

58. "Warning to Hospitals," *New York Times*, Mar. 27, 1906.

59. "Harburger Denounces Hospitals for Murder," *New York Times*, Mar. 7, 1906.

60. "Turn Sick Child Out of Hospital," *New York Times*, July 21, 1907.

61. "Hospital Patients Sent Away to Die," *New York Times*, Mar. 14, 1914.

62. Daniel M. Fox, *Power and Illness: The Failure and Future of American Health Policy* (Berkeley: University of California Press, 1993), p. 33; Sigmund S. Goldwater, "Crusading for the Chronically Sick," *Modern Hospital* 44, no. 5 (May 1935): 65; E. H. Lewinski-Corwin, *The Hospital Situation in Greater New York: Report of a Survey of Hospitals in New York City* (New York: Putnam's, 1924), p. 68.

63. Starr, *Social Transformation*, p. 157; see Mary C. Jarrett, *Chronic Illness in New York City*, 2 vols. (New York: Columbia University Press, 1933).

64. See New York City Cancer Hospital, *Twelfth Annual Report, 1896* (New York: Knickerbocker Press, 1896), p. 11.

65. Susan M. Wood, "Hospital Facilities for the Treatment of Cancer," *American Journal of Public Health* 20, no. 8 (Aug. 1930): 853.

66. "The Intern Remarks," American Life Histories: Manuscripts from the Federal Writers' Project, 1936–1940, AMLC.

67. S. S. Goldwater, *On Hospitals* (New York: Macmillan, 1947), p. 152.

68. E. M. Bluestone, "The 'Chronic' Has a Claim to Care and Cure in the 'Acute' General Hospital," *Modern Hospital* 63, no. 3 (Sept. 1944): 67–69.

69. See Fox, *Power and Illness.*

70. Letter of John A. Kingsbury, Jan. 19, 1915, file 785, box 75, Office of the Mayor, Mitchel Administration, MP; Frederick M. Dearborn, *The Metropolitan Hospital: A Chronicle of 62 Years* (New York: privately printed, 1937), p. 135.

71. Bluestone, "The 'Chronic' Has a Claim to Care and Cure," p. 67.

72. "Report of Conference in re. Proposed Tubercular Clinics, 1923," OD 1821H, pp. 18–19, LACBS.

73. Quoted in Frederic A. Washburn, *The Massachusetts General Hospital: Its Development, 1900–1935* (Boston: Houghton Mifflin, 1939), p. 88.

74. Margaret Caldwell Hilbert, "Nobody Asked You to Come," *Mount Sinai Alumnae News*, Mt. Sinai Hospital School of Nursing (Spring 1998): 3–4, in Mt. Sinai Archives, Mt. Sinai Hospital, New York.

75. See Judith Walzer Leavitt, *Brought to Bed: Child-Bearing in America, 1750–1950* (New York: Oxford University Press, 1986), pp. 179–87.

76. Even the rare article entitled "The Critically Ill Patient" (*Modern Hospital* [Sept. 1926]: 77–81) concentrated on postmortem care.

77. Helen Sheppard Dunlap, "A Nurse Recalls," in *Roosevelt Hospital*, p. 227.

78. Goldwater, "On Humanizing the Hospital," p. 543.

79. Albany Hospital, *Regulations and Routines for Resident Staff* (Albany: Albany Hospital, 1939), pp. 34–35; Hornsby and Schmidt, *The Modern Hospital*, p. 504.

80. The Presbyterian Hospital patient records (PHPR) provide an example.

81. M. J. Drummond to William J. Gaynor, Mar. 30, 1911, file 539, box 64, Office of the Mayor, Gaynor Administration, MP.

82. Memorial Hospital for the Treatment of Cancer and Allied Diseases, *Twenty-ninth Annual Report for the Year 1913* (New York: Knickerbocker Press, 1913), pp. 87–88.

83. Rosner, *Once Charitable Enterprise*, p. 77.

84. "Hospital Silent on a Woman's Death," *New York Times*, Feb. 9, 1900.

85. Quoted in Rosenberg, *Care of Strangers*, p. 310.

86. "Blames Hospital in Warner Death," *New York Times*, Nov. 5, 1910.

87. Stevens, *In Sickness and in Wealth*, p. 123.

88. *Report of the Board of Administrators of the Charity Hospital to the General Assembly of the State of Louisiana, 1883* (New Orleans: A. W. Hyatt, 1884), p. 9.

89. John E. Salvaggio, *New Orleans' Charity Hospital: A Story of Physicians, Politics, and Poverty* (Baton Rouge: Louisiana State University Press, 1992), p. 125.

90. See Harry F. Dowling, *City Hospitals: The Undercare of the Underprivileged* (Cambridge: Harvard University Press, 1982).

91. Lewinski-Corwin, *Hospital Situation in Greater New York*, p. 220.

92. "Los Angeles County Hospital," OD3706H, Report No. 21, LACBS.

93. Lewinski-Corwin, *Hospital Situation in Greater New York*, p. 220.

94. Walter Sands Mills, *The Tuberculosis Infirmary of the Metropolitan Hospital, Department of Public Charities, New York City* (New York: Martin B. Brown, 1908), p. 46.

95. Arthur Ames Bliss, *The Blockley Days* (n. p.: privately printed, 1916).

96. "Los Angeles County Hospital," OD3706H, Report No. 21, Mar. 2, 1914 (submitted to the board by Dr. Burt F. Howard, director of the State Bureau of Tuberculosis, Sacramento), p. 2, LACBS.

97. Edythe Tate Thompson to Los Angeles County Board of Supervisors, OD 5111H, Dec. 12, 1921, LACBS.

98. Quoted in Salvaggio, *New Orleans' Charity Hospital*, p. 125.

99. *Rules and Regulations of the City Hospital, Blackwell's Island*, Adopted by the Commissioner of Public Charities, Hon. John W. Keeler, November 1, 1901 (New York, 1901), p. 14, Library, New York Academy of Medicine.

100. New York City Charity Organization Society (COS) cases no. R684, box 251; R 958, box 266; R198, box 277; R995, box 277, CSS.

101. John A. Kingsbury to Edward S. McSweeny, June 30, 1914, file 684, box 75, Office of the Mayor, Mitchel Administration, MP.

102. M. L. Fleming, "Memorandum for Dr. Brannan," Mar. 1915, file 91, box 11, Office of the Mayor, Mitchel Administration, MP.

103. John W. Brannan to S. L. Martin, Feb. 6, 1917, and George O'Manlon, "Memorandum for Dr. Brannan," Feb. 6, 1917, both in file 90, box 11, Office of the Mayor, Mitchel Administration, MP. See also Sandra Opdycke, *No One Was Turned Away: The Role of Public Hospitals in New York City since 1900* (New York: Oxford, 1999), p. 57.

104. COS case no. R266, box 251, CSS.

105. Quoted in Regina Markell Morantz, "Introduction," in *In Her Own Words: Oral Histories of Women Physicians*, ed. Regina Markell Morantz, Cynthia Stodala Pomerleau, and Carol Hansen Fenichel (New Haven: Yale University Press, 1982), p. 29.

106. Harriett M. Bartlett, "Ida M. Cannon: Pioneer in Medical Social Work," *Social Service Review* 49, no. 2 (June 1975): 208–29.

107. Ibid., p. 214.

108. See, e.g., Ida M. Cannon, *Social Work in Hospitals: A Contribution to Progressive Medicine* (New York: Survey Associates, 1913); Social Service Bureau, Bellevue and Allied Hospitals, New York, *Report for the Year 1911* (New York: Bellevue and Allied Hospitals, 1911); Peter Bent Brigham Hospital, "Report of Social Service," *Seventh Annual Report for the Year 1920* (Cambridge, MA: The University Press, 1921), 34–41; Social Service Department, Massachusetts General Hospital, *Thirteenth Annual Report, January 1, 1918 to January 1, 1919* (Boston: Massachusetts General Hospital, 1919); Social Service Department, Barnes Hospital, St. Louis Children's Hospital, Washington University Dispensary, *Fourth and Fifth Annual Report* (St. Louis, 1916).

109. Richard C. Cabot, "A Plea for a Clinical Year in the Course of Theological Study," in *Adventures on the Borderlands of Ethics*, by Richard C. Cabot (New York: Harper & Brothers, 1925), pp. 1–22.

110. Ibid., p. 16.

111. "Chaplaincy, History," www.massgeneral.org/visitor/chaplain_history.htm, accessed Nov. 13, 2011.

112. Richard C. Cabot and Russell L. Dicks, *The Art of Ministering to the Sick* (New York: Macmillan, 1936), p. 298.

113. William Munk, *Euthanasia; or, Medical Treatment in Aid of an Easy Death* (New York: Arno, 1887).

114. W. Bruce Fye, "Active Euthanasia: An Historical Survey of Its Conceptual Origins and Introduction into Medical Thought," *Bulletin of the History of Medicine* 52 (1979): 492–502; Shai J. Lavi, *The Modern Art of Dying: A History of Euthanasia in the United States* (Princeton: Princeton University Press, 2005).

115. Munk, *Euthanasia*, p. 7.

116. Nic Hughes and David Clark, "'A Thoughtful and Experienced Physician': William Munk and the Care of the Dying in Late Victorian England," *Journal of Palliative Medicine* 7, no. 5 (2004): 708; see also Michael Bliss, *William Osler: A Life in Medicine* (New York: Oxford University Press, 1999).

117. See Paul S. Mueller, "William Osler's Study of the Act of Dying: An Analysis of the Original Data," *Journal of Medical Biography* 15, Suppl. 1 (2007): 60.

118. Bliss, *William Osler.*

119. Worcester, "Care of the Dying," pp. 200–224.

120. Edna L. Foley, *Visiting Nurse Manual* (Chicago: National Organization for Public Health Nursing, 1915), p. 31.

121. Quoted in Robert C. Wells, *Facing the "King of Terrors": Death and Society in an American Community, 1750–1990* (New York: Cambridge University Press, 2000), p. 181.

122. Ruth I. Peffly, "Monthly Report," Browning, MT, Apr. 1934, Records of the Bureau of Indian Affairs, RG 75, file E779, National Archives, Washington, DC.

123. Jacqueline H. Wolf, *Don't Kill Your Baby: Public Health and the Decline of Breastfeeding in the Nineteenth and Twentieth Centuries* (Columbus: Ohio State University Press, 2001), p. 125.

124. Nancy Tomes, *The Gospel of Germs: Men, Women, and the Microbe in American Life* (Cambridge: Harvard University Press, 1998), p. 184.

125. See Karen Halttunen, *Murder Most Foul: The Killer and the American Gothic Imagination* (Cambridge: Harvard University Press, 1998).

126. See Richard A. Meckel, *Save the Babies: American Public Health Reform and the Prevention of Infant Mortality, 1850–1929* (Baltimore: Johns Hopkins University Press, 1990).

127. Quoted in Wolfe, *Don't Kill Your Baby*, p. 44.

128. See John Duffy, "Social Impact of Disease in the Late Nineteenth Century," *Sickness and Health in America: Readings in the History of Medicine and Public Health*, ed. Judith Walzer Leavitt and Ronald Numbers (Madison: University of Wisconsin Press, 1997), pp. 418–25.

129. Quoted in Molly Ladd-Taylor, *Mother-Work: Women, Child Welfare, and the State, 1890–1930* (Urbana: University of Illinois Press, 1994), p. 33.

130. Julia C. Lathrop, "The Imperative Need of Safeguarding Maternity and Infancy," in *American Child Health Association, Tenth Annual Meeting* (Asheville, NC, 1919), p. 30.

131. Quoted in Wolf, *Don't Kill Your Baby*, p. 44.

132. Florence Kelley, "Children in the Cities," *National Municipal Review* 4, no. 14 (Apr. 1915): 197–203.

133. S. Josephine Baker, *Fighting for Life* (New York: Macmillan, 1939), p. 58.

134. Quoted in Ladd-Taylor, *Mother-Work*, p. 19.

135. Social Service Bureau, Bellevue and Allied Hospitals, *Report for the Year 1911*, pp. 19–20.

136. Luella Erion, "Francesco of Arizona," *Public Health Nurse* 13, no. 3 (Spring 1921): 315–34.

137. See Ladd-Taylor, *Mother-Work.* For British examples, see Ellen Ross, *Love and Toil: Motherhood in Outcast London, 1870–1918* (New York: Oxford University Press, 1993), and Julie-Marie Strange, *Death, Grief, and Poverty in Britain, 1870–1914* (New York: Cambridge University Press, 2005).

138. See Emily K. Abel, "Taking the Cure to the Poor: Patients' Responses to New York City's Tuberculosis Program, 1894–1918," *American Journal of Public Health* 87, no. 11 (Nov. 1997): 1808–15.

139. Ladd-Taylor, *Mother-Work*, p. 82.

140. Florence Kelley, *Notes of Sixty Years: The Autobiography of Florence Kelley*, ed. Kathryn Kish Sklar (Chicago: Charles H. Kerr, 1986), pp. 30–31.

141. Kathryn Kish Sklar, *Florence Kelley and the Nation's Work: The Rise of Women's Political Culture, 1830–1900* (New Haven: Yale University Press, 1995), p. 28.

Chapter 4 • Institutionalizing the Incurable

1. Minnie Patterson to Mary Post Zimmerman, Dec. 15, 1912, file 4, box 3, Zimmerman Family Papers, HL.

2. Louis I. Dublin, Edwin W. Kopf, and George H. Van Buren, *Cancer Mortality among Insured Wage Earners and Their Families: The Experience of the Metropolitan Life Insurance Company Industrial Department, 1911–1922* (New York: Metropolitan Life Insurance Co., 1925).

3. Quoted in Keith Wailoo, *How Cancer Crossed the Color Line* (New York: Oxford University Press, 2011), p. 13.

4. George St. J. Perrott, "The Problem of Chronic Disease," *Psychosomatic Medicine* 7 (Jan. 1, 1945): 22.

5. Ibid., p. 22.

6. Dublin, Kopf, and Van Buren, *Cancer Mortality*; Louis I. Dublin, "Decline of Tuberculosis: Present Death Rates and Outlook for the Future," *American Review of Tuberculosis* 43 (1941): 227–28; Richard Shryock, *National Tuberculosis Association 1904–1954: A Study of the Voluntary Health Movement in the United States* (New York: National Tuberculosis Association, 1957), pp. 302–3.

7. "Consumption Is a Preventable and Curable Disease," in Barbara Gutmann Rosenkrantz, *From Consumption to Tuberculosis: A Documentary History* (New York: Garland Publishing, 1994), p. 435.

8. Quoted in Katherine Ott, *Fevered Lives: Tuberculosis in American Culture since 1870* (Cambridge: Harvard University Press, 1996), p. 81.

9. Robert A. Aronowitz, *Unnatural History: Breast Cancer and American Society* (New York: Cambridge University Press, 2007); Kirsten Gardner, *Early Detection: Women, Cancer, and Awareness Campaigns in the Twentieth-Century United States* (Chapel Hill: University of North Carolina Press, 2006); Barron H. Lerner, *The Breast Cancer Wars: Fear, Hope, and the Pursuit of a Cure in Twentieth-Century America* (New York: Oxford University Press, 2001); James T. Patterson, *The Dread Disease: Cancer and Modern American Culture* (Cambridge: Harvard University Press, 1987), p. 113.

10. Isaac Levin, "The Cancer Problem and the Nurse," *American Journal of Nursing* 27, no. 2 (Feb. 1927), p. 88.

11. David Cantor, "Uncertain Enthusiasm: The American Cancer Society, Public Education, and the Problems of the Movie, 1921–1960," *Bulletin of the History of Medicine* 81, no. 1 (Spring 2007): 51.

12. W. Bruce Fye, *American Cardiology: The History of a Specialty and Its College* (Baltimore: Johns Hopkins University Press, 1996); quotation from p. 65.

13. See Aronowitz, *Unnatural History*; Cantor, "Uncertain Enthusiasm"; Gardiner, *Early Detection*; Lerner, *Breast Cancer Wars*.

14. Quoted in Fye, *American Cardiology*, p. 71.

15. Quoted in Wailoo, *How Cancer Crossed the Color Line*, p. 25.

16. Charles R. Grandy, "The Negro Consumptive," *Southern California Practitioner* 23, no. 6 (June 1908): 243, 245.

17. Violet Blanche Goldberg, "A Study of the Home Treatment of Tuberculosis Cases with Details of a Colony Plan in Los Angeles County and a Study of a Family Group of Ninety-Nine" (M.A. thesis, University of Southern California School of Social Work, Los Angeles, June 1939), p. 7.

18. See Wailoo, *How Cancer Crossed the Color Line*, p. 2.

19. Ernest A. Sweet, "Interstate Migration of Tuberculous Persons: Its Bearing on the Public Health, with Special Reference to the States of Texas and New Mexico," *Public Health Reports* (Apr. 23, 1915): 1240.

20. See Nancy Tomes, *The Gospel of Germs: Men, Women, and the Microbe in American Life* (Cambridge: Harvard University Press, 1998).

21. Emily K. Abel, *Tuberculosis and the Politics of Exclusion: A History of Public Health and Migration to Los Angeles* (New Brunswick, NJ.: Rutgers University Press, 2007); Sheila M. Rothman, *Living in the Shadow of Death: Tuberculosis and the Social Experience of Illness in American History* (New York: Basic Books, 1994).

22. Emily K. Abel, "Medicine and Morality: The Health Care Program of the New York Charity Organization Society," *Social Service Review* 71, no. 4 (Dec. 1997): 634–51.

23. *Tuberculosis Families in Their Homes: A Case Study* (New York: Association of Tuberculosis Clinics and the Committee on the Prevention of Tuberculosis of the Charity Organization Society of the City of New York, 1916), p. 16.

24. Grace Abbott, "The Social Security Act and Relief," *University of Chicago Law Review* 4 (Dec. 1936): 67, quoting the *Report to the President of the Committee on Economic Security* (Washington, DC: U.S. Government Printing Office, 1935).

25. Ernest A. Sweet, "Interstate Migration of Tuberculous Persons: Its Bearing on the Public Health, with Special Reference to the States of Texas and New Mexico," *Public Health Reports* (Apr. 16, 1915): 1150.

26. Charles Dwight Willard to Samuel Willard, Charles Dwight Willard Collection, HL. The letter is dated 1909, but its contents indicate that it should be sometime in spring 1910.

27. Emily K. Abel, *Suffering in the Land of Sunshine: A Los Angeles Illness Narrative* (New Brunswick, NJ: Rutgers University Press, 2006).

28. Petitions to the Board of Supervisors, Aug. 20, 1908, file OD949H, LACBS.

29. Susan M. Schweik, *The Ugly Laws: Disability in Public* (New York: New York University Press, 2009); see also Adrienne Phelps Coco, "Diseased, Maimed, Mutilated: Categorizations of Disability and an Ugly Law in Late Nineteenth-Century Chicago," *Journal of Social History* 44, no. 1 (Fall 2010): 23–37.

30. Quoted in Schweik, *The Ugly Laws*, p. 33.

31. Quoted in Rothman, *Living in the Shadow of Death*, pp. 196–97.

32. Godias J. Drolet, "Tuberculosis Hospitalization in the United States: Results, Types of Cases, Facilities, and Costs," *American Review of Tuberculosis*, Dec. 1926, pp. 600–601, 605.

33. Hermann M. Biggs, *The Administrative Control of Tuberculosis* (New York: New York City Department of Health, 1909), p. 21.

34. Charles-Edward Armory Winslow, *The Life of Hermann Biggs, M.D., D.Sc., LL.D., Physician and Statesman of the Public Health* (Philadelphia: Lea & Febiger, 1929), p. 198.

35. Quoted in Barron H. Lerner, *Contagion and Confinement: Controlling Tuberculosis along the Skid Road* (Baltimore: Johns Hopkins University Press, 1998), p. 28.

36. R. L. Cunningham, "The Barlow Sanatorium, Los Angeles, Cal.: Report on Sixty Cases Discharged during the Year, September, 1908–September, 1909," *Southern California Practitioner* 24, no. 12 (Dec. 1909): 607.

37. See Barbara Bates, *Bargaining for Life: A Social History of Tuberculosis, 1876–1938* (Philadelphia: University of Pennsylvania Press, 1992).

38. *Eighteenth Annual Report of Montefiore, a Hospital for Chronic Invalids, and Country Sanitarium for Consumptives, November 1902* (New York: Oppenheimer Printing Co., 1902), p. 55.

39. *Twenty-Fourth Annual Report of Montefiore, a Hospital for Chronic Invalids, and Country Sanitarium for Consumptives, November 1908* (New York: Oppenheimer Printing Co., 1908), pp. 61–62.

40. Emil Frankel and Helen E. Heyer, "Tuberculous Patients in New Jersey Sanatoriums," *American Journal of Public Health* 20 (1930): 976.

41. *Surgeon Errant: The Life and Writings of William Henry Bucher, 1874–1934*, ed. Emil Bogen (Los Angeles: Angelus Press, 1935), p. 139.

42. "Tuberculosis—Monthly Reports" (typewritten reports by Edythe Tate-Thompson, director of the Bureau of Tuberculosis), Jan. 1933, California State Archives, Sacramento.

43. Frankel and Heyer, "Tuberculous Patients," p. 971.

44. Cunningham, "Barlow Sanatorium," p. 607.

45. Bogen, *Surgeon Errant*, p. 122.

46. Quoted in Alan Bauman, "From Tuberculosis Sanatorium to Medical Center: The History of Olive View Medical Center, 1920–1989" (M.A. thesis, University of California, Santa Barbara, 1989), p. 41.

47. Bates, *Bargaining for Life*, p. 270.

48. Drolet, "Tuberculosis Hospitalization," p. 609.

49. Bates, *Bargaining for Life*, p. 54; Georgina D. Feldberg, *Disease and Class: Tuberculosis and the Shaping of Modern North American Society* (New Brunswick, NJ: Rutgers University Press, 1995), pp. 101–2; Richard Sucre, "The Great White Plague: The Culture of Death and the Tuberculosis Sanatorium," www.faculty.virgina.edu/blueridgesanatorium/death.htm, accessed Oct. 20, 2007.

50. See Sucre, "Great White Plague."

51. Ibid.

52. Rothman, *Living in the Shadow of Death*, pp. 238–43; Sucre, "Great White Plague."

53. Sadie Fuller Seagrave, *Saints' Rest* (St. Louis: C. V. Mosby, 1918), p. 47.

54. Betty MacDonald, *The Plague and I* (Philadelphia: J. B. Lippincott, 1948), p. 151.

55. Marshall McClintock, *We Take to Bed* (New York: Jonathan Cape & Harrison Smith, 1931), p. 181.

56. See, e.g., MacDonald, *Plague and I*, p. 141; *Madonna Swan: A Lakota Woman's Story*, as told through Mark St. Pierre (Norman: University of Oklahoma Press, 1992), pp. 79–80.

57. Will Ross, *I Wanted to Live: An Autobiography* (Milwaukee: Wisconsin Anti-Tuberculosis Assoc., 1953), p. 84; Rothman, *Living in the Shadow of Death*, pp. 238–43.

58. Ross, *I Wanted to Live*, p. 95.

59. MacDonald, *Plague and I*, p. 141.

60. Ross, *I Wanted to Live*, p. 115.

61. Mary Sewall Gardner, *Public Health Nursing*, 2nd ed. (New York: Macmillan, 1931), p. 275.

62. Charity Organization Society case no. R52, box 240, CSS.

63. National Tuberculosis Association, *A Directory of Sanatoria, Hospitals, Day Camps and Preventoria for the Treatment of Tuberculosis in the United States* (New York: National Tuberculosis Assoc., 1923), p. 16.

64. Interview with Mrs. Emilia Castañeda de Valenciana by Christine Valenciana, Sept. 8, 1971, Oral History Program, California State University, Fullerton.

65. Ibid.

66. New York Skin and Cancer Hospital, *First Annual Report* (New York, 1884), p. 10. All reports of this hospital were published by the hospital itself.

67. New York Skin and Cancer Hospital, *Second Annual Report* (New York, 1885), pp. 8–9.

68. New York Skin and Cancer Hospital, *Twenty-fifth Annual Report* (New York, 1907), pp. 13–14, 18.

69. The phrase comes from the New York Skin and Cancer Hospital, *Eleventh and Twelfth Annual Reports, 1894–1895* (New York, 1895), p. 20. See also Harry C. Saltzstein, "The Average Treatment of Cancer," *JAMA* 91, no. 7 (Aug. 18, 1928): 465–70; Susan M. Wood, "Hospital Facilities for the Treatment of Cancer," *American Journal of Public Health* 20, no. 8 (Aug. 1930): 849–61.

70. New York Cancer Hospital, *Fourth Annual Report for the Year 1888* (New York: G. P. Putnam's, 1888), p. 11.

71. New York Cancer Hospital, *Seventh Annual Report for the Year 1891* (New York: G. P. Putnam's, 1891), p. 11; *Thirteenth Annual Report for the Year 1897* (New York: Knickerbocker Press, 1897), p. 16; and *Fourteenth Annual Report for the Year 1898* (New York: Knickerbocker, 1898), p. 15.

72. Memorial Hospital for the Treatment of Cancer and Allied Diseases, *Twenty-ninth Annual Report for the Year 1913* (New York: Knickerbocker, 1913), pp. 86–87.

73. General Memorial Hospital, *Twenty-first Annual Report for the Year 1905* (New York: Knickerbocker, 1905), pp. 17–18; Memorial Hospital for the Treatment of Cancer and Allied Diseases, *Thirty-first Annual Report for the Year 1915* (New York: Knickerbocker, 1915), pp. 37–38.

74. Memorial Hospital for the Treatment of Cancer and Allied Diseases, *Thirty-fourth Annual Report for the Years 1918–1919–1920* (New York: Knickerbocker, 1920), p. 18, and *Thirty-fifth Report for 1921–22* (New York: Knickerbocker, 1922), p. 45.

75. William H. Livingston, "Social Service for Cancer Patients," *Modern Hospital* 23, no. 6 (Dec. 1924): 553.

76. Memorial Hospital, *Thirty-fourth Annual Report*, p. 18. See Mary C. Jarrett, *Chronic Illness in New York City* (New York: Columbia University Press, 1933), vol. 1.

77. Jarrett, *Chronic Illness*, 1:179.

78. Ibid., pp. 179, 249, 46.

79. See the following web pages, accessed Nov. 2010: "Beechwood History," www.beechwoodhome.org/BeechwoodHistory.aspx; "Historical Perspective, Cedars-Sinai Medical

Center, Los Angeles," cedars-sinai.edu/About-Us/History/Documents/HistPersp 703.pdf; "The Boston Home Growing and Changing with the Times: Interview," http://findarticles .com/p/articles/mi_m3830; "Crozer Home for Incurables," www.oldchesterpa.com/crozer _home.htm; "Important Dates in Lawrenceville History," www.lhs 15201.org/articles_b .asp?ID=19; "The Storrs Society," www.calvaryhospital.org/site/pp.asp?c; "A Timeline of Benevolent Giving and Humane Goodness, 1851–1900," www.msa.md.gov/msa/speccol /photos/philanthropy/html/timeline2.htm; "The Virginia Home: A Rich Tradition of Caring," www.thevirginiahome.org/history.aspx; Harry J. Warthen, "Medicine and Shockoe Hill: The Medical College of Virginia's Records of Service," http://richmondthenand now .com/Newspaper-Articles; George Zepin, "Intermunicipal Co-operation in Charitable Activities," *Proceedings of the National Conference of Jewish Charities*, pp. 69–74, www .policyarchive.org/handle/10207bitstreams/9664. Also see Ruth Everett, "American House of Calvary," *Catholic World* 75 (Aug. 1902): 628–33; Darlene Roth, "Feminine Marks on the Landscape: An Atlanta Inventory," *Journal of American Culture* 3, no. 4 (Winter 1980): 673–85.

80. Katherine Burton, *Sorrow Built a Bridge: A Daughter of Hawthorne* (London: Longmans, Green, 1937); Theodore Maynard, *A Fire Was Lighted: The Life of Rose Hawthorne Lathrop* (Milwaukee: Bruce Publishing, 1948); Patricia Dunlavy Valenti, *To Myself a Stranger: A Biography of Rose Hawthorne Lathrop* (Baton Rouge: Louisiana State University Press, 1991); James J. Walsh, *Mother Alphonsa: Rose Hawthorne Lathrop* (New York: Macmillan, 1930).

81. See, e.g., Burton, *Sorrow*; Walsh, *Mother Alphonsa*.

82. Mother Mary Alphonsa to Clifford Smyth, May 26, 1922, folder 3, box 2, Hawthorne Family Papers, Department of Special Collections, Green Library, Stanford University, Stanford, CA.

83. Flannery O'Connor, introduction to *Mission Fulfilled: Heartwarming Story of Courage*, by Sister M. Evangelist (New York: Dell, 1962), pp. 22, 11.

84. See Bernadette McCauley, *Who Shall Take Care of Our Sick? Roman Catholic Sisters and the Development of Catholic Hospitals in New York City* (Baltimore: Johns Hopkins University Press, 2005), pp. 16–19.

85. Mother Alphonsa to Clifford Smyth, Apr. 8, 1919, folder 1, box 2, Hawthorne Family Papers.

86. Lorraine Gracey, "Home for Indigent Cancer Patients," *New York Times*, Dec. 5, 1979; Maynard, *Fire Was Lighted*; Valenti, *To Myself a Stranger*; Walsh, *Mother Alphonsa*.

87. Valenti, *To Myself a Stranger*; Walsh, *Mother Alphonsa*.

88. Rose Hawthorne Lathrop, "Will Be a Hospital Home" (letter to the editor), *New York Times*, July 27, 1897.

89. "Two Homes That Smile as One," *Report of the Dominican Cancer Homes for the Destitute in New York and Westchester Co., New York*, Oct. 1915–Oct. 1916, p. 3.

90. Quoted in Joy Buck, "Reweaving a Tapestry of Care: Religion, Nursing, and the Meaning of Hospice, 1945–1978," *Nursing History Review* 15 (2007): 116.

91. Rose Hawthorne Lathrop, "Appeal for Cancer Patients" (letter to the editor), *New York Times*, Nov. 13, 1897.

92. Mother Mary Alphonsa Lathrop, "Appeal for Aid for Cancerous Poor" (letter to the editor), *New York Times*, Oct. 5, 1902.

93. Quoted in "Nursing Cancer Patients," *New York Times*, Dec. 23, 1900; Valenti, *To Myself a Stranger*, p. 154.

94. Quoted in Walsh, *Mother Alphonsa*, pp. 83–85.

95. Quoted in Valenti, *To Myself a Stranger*, p. 141.

96. Ibid., pp. 142–43.

97. Marynard, *A Fire Was Lighted*, p. 353.

98. Julian J. Hawthorne, "Daughter of Hawthorne," *Atlantic Monthly* 142 (Sept. 1928): 375.

99. Carrie Webster Nichols, "Mrs. Lathrop's Work" (letter to the editor), *New York Times*, Mar. 15, 1899.

100. William Ian Miller, *The Anatomy of Disgust* (Cambridge: Harvard University Press, 1998), pp. 194, 197.

101. Mother Alphonsa to Sister Rose, June 8, 1914, RHL.

102. Susan E. Lederer, *Subjected to Science: Human Experimentation in America before the Second World War* (Baltimore: Johns Hopkins University Press, 1995).

103. Mother Alphonsa to Sister Rose, Aug. 8 and Dec. 15, 1902, RHL.

104. Mother M. Rose Huber, "Report of a Case at St. Rose's Free Home," *Report of the Dominican Cancer Homes for the Destitute*, p. 9.

105. Cited in Buck, "Reweaving a Tapestry," p. 121.

106. Quoted in Buck, "Reweaving a Tapestry," p. 121. The success of the hospice movement that Saunders inspired gradually undermined the need for homes for the incurable. Although the Dominican Sisters of Hawthorne recently established a new facility in Kisumu, Kenya, it has closed many of the U.S. homes.

107. Editorial, *Modern Hospital*, Dec. 1931, p. 103.

108. Dorothy Levenson, *Montefiore: The Hospital as Social Instrument* (New York: Farrar, Straus & Giroux, 1984), p. 177.

109. Ibid., p. 20.

110. Levenson, *Montefiore*, p. 15.

111. *Annual Report of the Directors of the Montefiore Home for Chronic Invalids, November 1886* (New York: Stettiner, Lambert, 1886), p. 13.

112. *Annual Report of the Directors of the Montefiore Home for Chronic Invalids, November 1887* (New York: LeLeeu, Oppenheimer & Myers, 1887), p. 14.

113. "Montefiore Medical Center Milestones," www.montefiore.org.

114. Levenson, *Montefiore*, p. 87.

115. Milton Bracker, *Montefiore Hospital for Chronic Diseases, New York: Fiftieth Anniversary* (New York: privately printed, 1934), p. 5.

116. See Patricia Spain Ward, *Simon Baruch: Rebel in the Ranks of Medicine, 1840–1921* (Tuscaloosa: University of Alabama Press, 1994).

117. Quoted in Levenson, *Montefiore*, p. 29; *Annual Report of the Directors of the Montefiore Home for Chronic Invalids, November 1886*, pp. 16–17.

118. Levenson, *Montefiore*; Ward, *Simon Baruch*.

119. Quoted in Levenson, *Montefiore*, pp. 92–93.

120. Ibid., p. 93.

121. *Thirty-seventh Annual Report of the Montefiore Hospital for Chronic Diseases, November 1921* (New York: William C. Popper, 1921), p. 13.

122. Bracker, *Montefiore*, p. 33.

123. Jane Pacht Brickman, "Ernst P. Boas (1891–1955)," *Journal of Public Health Policy* 20, no. 3 (1999): 348–55.

124. *Eighteenth Annual Report of the Directors of the Montefiore Home, A Hospital for Chronic Invalids and Country Sanitarium for Consumptives, November 1902* (New York: Press of the Oppenheimer Printing, 1902), p. 14.

125. *Thirtieth Annual Report of Montefiore Home and Hospital for Chronic Invalids, November, 1914* (New York: William C. Popper, 1914), pp. 11–12.

126. Jarrett, *Chronic Illness*, 2:46–47.

127. *Thirty-seventh Annual Report of Montefiore Hospital for Chronic Diseases, November 1921*, p. 50.

128. *Forty-seventh Annual Report of the Trustees of the Montefiore Hospital for Chronic Diseases, for the Year 1931* (New York: William C. Popper, 1931), p. 31.

129. Harry F. Dowling, *City Hospitals: The Undercare of the Underprivileged* (Cambridge: Harvard University Press, 1982), p. 153.

130. Charles E. Rosenberg, *The Care of Strangers: The Rise of America's Hospital System* (New York: Basic Books, 1987).

131. *Majority and Minority Reports on Investigation of Boston Almshouse and Hospital at Long Island* (Boston: Municipal Printing Office, 1904). When Bellevue and its affiliated hospitals in New York—Fordham, Harlem, and Gouverneur—were placed under a separate administrative board in the early twentieth century, the chronic care hospitals on Blackwell's Island remained under the control of the Department of Public Charities.

132. *Sixth Annual Report of Pauper Institutions of the City of Boston for the Year Ending Jan. 31, 1903* (Boston: Municipal Printing Office, 1903), p. 17.

133. Walter Sands Mills, *The Tuberculosis Infirmary of the Metropolitan Hospital, Department of Public Charities, New York City* (New York: Martin B. Brown, 1908), p. 15.

134. "Annual Report of the Tuberculosis Infirmary for the Year Ending December 31, 1903," in *Annual Report of the Department of Public Charities of the City of New York, 1903* (New York: Martin B. Brown, 1904), p. 178.

135. "Death Rate 96 Per Cent," *New York Times*, Dec. 29, 1897; "Care of the City's Poor" (letter to the editor from "A Missionary"), *New York Times*, Oct. 16, 1903.

136. "Mustn't Shift Patients," *New York Times*, Sept. 28, 1906; Frederick M. Dearborn, *The Metropolitan Hospital: A Chronicle of Sixty-two Years* (New York: privately printed, 1937), pp. 95–96.

137. Jarrett, *Chronic Illness*, 1:211. That change was made possible by the erection of the Queensboro Bridge, enabling ambulances to drive directly to both hospitals.

138. Dearborn, *The Metropolitan Hospital*, p. 135.

139. John A. Kingsbury to John P. Mitchel, Jan. 19, 1915, file 785, box 75,Office of the Mayor, Mitchel Administration, MP.

140. *Investigation of Boston Almshouse and Hospital*, pp. 4–5

141. "Crowding Patients in City Hospital," *New York Times*, Feb. 9, 1908.

142. J. F. to H. M. Johnson, n.d., filed with Charity Organization Society case no. R966, box 278, CSS.

143. John A. Kingsbury to John P. Mitchel, Jan. 22, 1915, file 785, box 75, Office of the Mayor, Mitchel Administration, MP.

144. Michael J. Drummond to Mayor William J. Gaynor, Mar. 20, 1911, file 539, box 64, Office of the Mayor, Gaynor Administration, MP.

145. "City's Sick Poor Seem Badly Aided," *New York Times*, Oct. 26, 1910.

146. John A. Kingsbury to the Board of Aldermen, City Hall, New York City, Feb. 11, 1915, file 785, box 75, Office of the Mayor, Mitchel Administration, MP.

147. Quotations from *Investigation of Boston Almshouse and Hospital*, pp. 1143 and 1112.

148. *Annual Report of the Pauper Institutions Department for the Year Ending Jan. 31, 1923* (Boston: Municipal Printing Office, 1923), p. 7.

149. Charity Organization Society case no. R693, box 268, CSS.

150. Charity Organization Society case no. R864, box 251, CSS.

151. S. S. Goldwater, "Crusading for the Chronically Sick," *Modern Hospital* 44, no. 5 (May 1935): 67.

152. See Michael B. Katz, *In the Shadow of the Poorhouse: A Social History of Welfare in America* (New York: Basic Books, 1986).

153. Helen W. Munson, "The Care of the Sick in Almshouses," *American Journal of Nursing* 30, no. 10 (Oct. 1930): 1228.

154. Ibid., p. 1229.

155. Ibid.

156. Ibid., p. 1230.

157. Carole Haber and Brian Gratton, "Old Age, Public Welfare, and Race: The Case of Charleston, South Carolina, 1800–1949," *Journal of Social History* 21, no. 2 (Winter 1987): 263–79.

158. Ibid., p. 269.

159. Ibid., p. 270.

160. Ibid., p. 268.

161. Homer Folks, "Disease and Dependence," *Charities* 10, no. 20 (May 1903): 499–500. See Carole Haber, *Beyond Sixty-Five: The Dilemma of Old Age in America's Past* (New York: Cambridge University Press, 1985), p. 82.

162. John A. Kingsbury to John P. Mitchel, Feb. 23, 1915, file 785, box 75, Office of the Mayor, Mitchel Administration, MP.

163. Mrs. Edmund G. Wendt to John P. Mitchel, September 15, 1916, file 793, box 76, Office of the Mayor, Mitchel Administration, MP.

164. Jarrett, *Chronic Illness*, 1:53.

165. "Der Moshev Zekeinim D'Harlem," *Jewish American*, July 1929; reprint available from the Hebrew Home at Riverdale, New York.

166. See N. Sue Weiler, "Religion, Ethnicity, and the Development of Private Homes for the Aged," *Journal of American Ethnic History* 12, no. 1 (Fall 1992): 64–90.

167. Carole Haber, "The Old Folks at Home: The Development of Institutionalized Care for the Aged in Nineteenth-Century Philadelphia," *Pennsylvania Magazine of History and Biography* 101, no. 2 (April 1977): 253; Weiler, "Religion, Ethnicity," p. 79; Daniel E. Lage, "'From the Blossom of Health, to the Paleness of Death': Transforming Aging, Illness, and Caregiving at the Boston Home for Aged Men, 1919–2003" (senior honors thesis, Department of History of Science, Harvard University, 2011).

168. Haber, "Old Folks at Home," pp. 253–54.

169. Quoted in Lage, "From the Blossom of Health," p. 2.

170. Jarrett, *Chronic Illness* 1:44, 107, 108.

171. A. E. Benjamin, "An Historical Perspective on Home Care Policy," *Milbank Quarterly* 71, no. 1 (1993): 129–66.

172. William Seaman Bainbridge, *The Cancer Problem* (New York: Macmillan, 1914), p. 434.

173. Ernst P. Boas, *The Unseen Plague: Chronic Disease* (New York: J. J. Augustin, 1940), p. 30.

174. "Report of the Commission on Relief of the Committee on the Prevention of Tuberculosis," Feb. 14, 1908, box 109, CSS.

175. Bainbridge, *The Cancer Problem*, p. 434; see also LeRoy Broun, "The Saving of Life in Cancer," *American Journal of Nursing* 25 (Mar. 1925): 194–98.

176. See Emily K. Abel, *Hearts of Wisdom: American Women Caring for Kin, 1850–1940* (Cambridge: Harvard University Press, 2000).

177. Emily K. Abel, "Taking the Cure to the Poor: Patients' Responses to New York City's Tuberculosis Program, 1894 to 1918," *American Journal of Public Health* 87, no. 11 (Nov. 1997): 1808–15.

178. Diary of Martha Shaw Farnsworth, Feb. 19, 1893, Martha Farnsworth Collection, Kansas State Historical Society, Topeka, KS.

179. The names in this paragraph are pseudonyms.

180. Brattleboro Cases, file 28, box 73, TTT.

181. Ibid.

182. Minnie Radamacher to Mary Post Zimmerman, June 5, 1915, file 12, box 3, Zimmerman Family Papers, HL.

183. See, e.g., Mildred C. J. Pfeiffer and Eloise M. Lemon, "A Pilot Study in the Home Care of Terminal Cancer Patients," *American Journal of Public Health* 43 (July 1953): 909–14; Prudence I. Priest and Virginia H. McCann, "Home Care for Mrs. Murphy," *American Journal of Nursing* 57, no. 12 (Dec. 1957): 1578–80; Sidney Shindell, "A Method of Home Care for Prolonged Illness," *Public Health Reports* 65, no. 20 (May 19, 1950): 651–60; *A Study of Selected Home Care Programs*, A Joint Project of the Public Health Service and the Commission on Chronic Illness, Public Health Monograph no. 35 (Washington, DC: U.S. Government Printing Office, 1955); John D. Thompson, "Nursing Service in a Home Care Program," *American Journal of Nursing* 51, no. 4 (April 1951): 233–35. See also Benjamin, "An Historical Perspective on Home Care Policy," 132–33; Eileen Boris and Jennifer Klein, *Caring for America: Home Health Workers in the Shadow of the Welfare State* (New York: Oxford University Press, 2012), p. 61.

184. Charity Organization Society case no. R602, box 263, CSS.

185. Ibid., case no. R806, box 273, CSS.

186. Ibid., case no. R2072, box 286, CSS.

187. Pseudonyms are used in this and the following paragraph.

188. Brattleboro cases, file 74, box 24; file 4, box 68; file 32, box 66; file 20, box 68; file 27, box 66, TTT.

189. Brattleboro cases, file 18, box 66, TTT.

190. *Twenty-second Annual Report of the Directors of the Montefiore Home, a Hospital for Chronic Invalids, and Country Sanitarium for Consumptives, November 1906* (New York: Oppenheimer Printing, 1906), p. 54.

191. Committee on the Costs of Medical Care, *Medical Care for the American People* (Chicago: University of Chicago Press, 1932).

192. Daniel M. Fox, "Policy and Epidemiology: Financing Health Services for the Chronically Ill and Disabled, 1930–1990," *Milbank Quarterly* 67, Suppl. 2, pt. 2 (1989):

257–87; Daniel M. Fox, *Power and Illness: The Failure and Future of American Health Policy* (Berkeley: University of California Press, 1993).

193. Fox, "Policy and Epidemiology" and *Power and Illness.*

194. Jarrett, *Chronic Illness,* 1:4. See also Fox, "Policy and Epidemiology."

195. Jarrett, *Chronic Illness,* 1:2.

196. Boas, *Unseen Plague,* p. 37.

197. Ibid., p. 30.

Chapter 5 • "All Our Dread and Apprehension"

1. Dushkin Diary, Dec. 6, 1959, DSD. On Scannell, see Earle Wayne Wilkins, "J. Gordon Scannell, MD (1914–2002)," *Journal of Thoracic and Cardiovascular Surgery* 125 (2003): 1–2.

2. On the development of megavolt radiation therapy, see *Final Report of the Advisory Committee on Human Radiation Experiments* (New York: Oxford University Press, 1996), p. 230. On actinomycin D, see L. C. Vining and S. A. Waksman, "Paper Chromatographic Identification of the Actinomycins," *Science* 120 (Sept. 3, 1954): 339; W. J. Burdette, "Alteration of Mutation Frequency by Treatment with Actinomycin D," *Science* 133 (Jan. 6, 1961): 140; E. Reich, R. M. Franklin, A. J. Shatkin, and E. L. Tatum, "Effect of Actinomycin D on Cellular Nucleic Acid Synthesis and Virus Production," *Science* 134 (Aug. 25, 1961): 5556–57.

3. Dushkin Diary, Mar. 27, 1960, DSD.

4. Ibid.

5. Dushkin Diary, Feb. 27, 1961.

6. "We Could Cure Cancer Now!" *Woman's Home Companion,* Nov. 1946, pp. 35, 176.

7. Richard Carter, *Breakthrough: The Saga of Jonas Salk* (New York: Trident Press, 1966), p. 1.

8. Quoted in David M. Oshinsky, *Polio, An American Story: The Crusade That Mobilized the Nation against the 20th Century's Most Feared Disease* (New York: Oxford University Press, 2005), p. 203.

9. Paul Starr, *The Social Transformation of American Medicine: The Rise of a Sovereign Profession and the Making of a Vast Industry* (New York: Basic Books, 1982), p. 346.

10. Waksman is discussed and quoted in Sheila M. Rothman, *Living in the Shadow of Death: Tuberculosis and the Social Experience of Illness in American History* (New York: Basic Books, 1994), p. 248.

11. "Leading Causes of Death," *Public Health Reports* 67, no. 1 (Jan. 1952): 92.

12. Quoted in W. Bruce Fye, *American Cardiology: The History of a Specialty and Its College* (Baltimore: Johns Hopkins University Press, 1996), p. 113.

13. Ibid., pp. 112–13.

14. "Cancer Fighter Cornelius P. Rhoads," *Time,* June 27, 1949.

15. Quoted in "Mayor Dedicates City Cancer Unit," *New York Times,* Aug. 24, 1950.

16. *Hygeia,* Dec. 1947, p. 916.

17. See Barron H. Lerner, *The Breast Cancer Wars: Fear, Hope, and the Pursuit of a Cure in Twentieth-Century America* (New York: Oxford University Press, 2001), p. 137; James T. Patterson, *The Dread Disease: Cancer and Modern American Culture* (Cambridge: Harvard University Press, 1987), pp. 195–98.

18. Patterson, *Dread Disease,* p. 196.

19. "Doctor Foresees Cancer Penicillin," *New York Times,* Oct. 3, 1953.

20. Patterson, *Dread Disease,* p. 154; "Progress Report," *Newsweek,* Jan. 20, 1958, p. 51.

21. "Cancer—On the Brink of Breakthroughs," *Life*, May 5, 1958, p. 102.

22. Henry Schacht, "Cancer and the Atom," *Harper's Magazine*, Aug. 1949, p. 83.

23. See, e.g., Walter C. Alvarez, "Search for a Cure for Cancer," *Good Housekeeping*, Apr. 1958, pp. 92, 248–50.

24. Alex and Julia Dushkin to Dorothy and David Dushkin, June 14, 1960, folder 5, box 1, DSD.

25. Alex and Julia Dushkin to Dorothy, David, and Nadia Dushkin, Nov. 26, 1960, folder 5, box 1, DSD.

26. Susan Allen Toth, *Ivy Days: Making My Way Out East* (New York: Ballantine, 1984), pp. 77–78.

27. William D. Kelly and Stanley R. Friesen, "Do Cancer Patients Want to Be Told?" *Surgery* 27, no. 6 (1950): 822.

28. Patterson, *Dread Disease*. Cf. Barbara Clow, *Negotiating Disease: Power and Cancer Care, 1900–1950* (Montreal: McGill-Queen's University Press, 2001).

29. See Isaac F. Marcosson, "Cured Cancer Club," *Hygeia*, Aug. 1939, pp. 694–96; "Cancer Contest Winners," *Hygeia*, Jan. 1941, pp. 25–27, 66–67.

30. Burton H. Wolfe, "The Cured Cancer Club: Now an International Institution," *Today's Health*, Jan. 1956, p. 32.

31. Terese Lasser, "I Had Breast Cancer," *Coronet*, Apr. 1954, p. 109.

32. See Susan E. Cayleff, *Babe: The Life and Legend of Babe Didrikson Zaharias* (Urbana: University of Illinois Press, 1995).

33. Q. Reynolds, "The Girl Who Lived Again," *Reader's Digest*, Oct. 1954, p. 55, quoted in Cayleff, *Zaharias*, pp. 234–35.

34. Quoted in Cayleff, *Zaharias*, p. 220.

35. Ibid., p. 239.

36. Stephen Humphrey Bogart, *Bogart: In Search of My Father* (New York: Dutton, 1995), p. 290.

37. William T. Fitts, Jr., and I. S. Ravdin, "What Philadelphia Physicians Tell Patients with Cancer," *JAMA* 153, no. 10 (1953): 901–4; Donald Oken, "What to Tell Cancer Patients: A Study of Medical Attitudes," *JAMA* 175, no. 13 (1961): 86–94. Cf. F. G. H. Maloney, "Should a Patient Be Told He Has Cancer?" *Wisconsin Medical Journal* 53 (1954): 541. See also Mary Ann Krisman-Scott, "An Historical Analysis of Disclosure of Terminal Status," *Journal of Nursing Scholarship*, First Quarter 2000, p. 50.

38. Oken, "What to Tell Cancer Patients."

39. Ibid.

40. See Ellen Herman, *Kinship by Design: A History of Adoption in the Modern United States* (Chicago: University of Chicago Press, 2008).

41. See Peter Galison, "Removing Knowledge," *Critical Inquiry* 31, no. 1 (2004): 229–43; Patrick Radden Keefe, *Chatter: Dispatches from the Secret World of Global Eavesdropping* (New York: Random House, 2005); Daniel Patrick Moynihan, *Secrecy: The American Experience* (New Haven: Yale University Press, 1989).

42. Oken, "What to Tell Cancer Patients," p. 1123.

43. Peter De Vries, *The Blood of the Lamb* (Chicago: University of Chicago Press, 1961), p. 169.

44. John Gunther, *Death Be Not Proud* (New York: Harper and Row, 1949), p. 46. For an excellent historical analysis of this book, see Gretchen Krueger, *Hope and Suffering:*

Children, Cancer, and the Paradox of Experimental Medicine (Baltimore: Johns Hopkins University Press, 2008).

45. PHPR #76015.

46. PHPR #726014.

47. PHPR #494039.

48. Bogart, *Bogart*, p. 280.

49. Laura Furman, *Ordinary Paradise* (Houston: Winedale, 1998).

50. Vries, *Blood of the Lamb*, p. 202.

51. Katinka Loeser, "Whose Little Girl Are You?" *New Yorker*, July 7, 1962, pp. 26–31.

52. Dushkin Diary, Apr. 11, 1960, and Jan. 5, 1962, DSD.

53. Robert W. Creamer, *Babe: The Legend Comes to Life* (New York: Simon and Schuster, 1974), p. 423.

54. Gunther, *Death Be Not Proud*, p. 46.

55. Maloney, "Should a Patient Be Told?" p. 541.

56. Furman, *Ordinary Paradise*, p. 61.

57. *Terminal Care for Cancer Patients: A Survey of the Facilities and Services Available and Needed for the Terminal Care of Cancer Patients in the Chicago Area* (Chicago: Central Service of the Chronically Ill of the Institute of Medicine of Chicago, 1950), p. 74.

58. Talcott Parsons, *The Social System* (New York: Free Press, 1951).

59. Barney G. Glaser and Anselm L. Strauss, *Awareness of Dying* (New Brunswick, NJ: Transaction, 1965), pp. 86, 99.

60. "Margaret Harding" is a pseudonym.

61. PHPR #759825.

62. Burton H. Wolfe, "The Cured Cancer Club," *Today's Health*, Dec. 1953, p. 25.

63. Wolfe, "Cured Cancer Club: Now an International Institution," pp. 32–35.

64. Dushkin Diary, Feb. 27, 1961, DSD.

65. Gunther, *Death Be Not Proud*, p. 3.

66. Dushkin Diary, Jan. 5, 1962, DSD.

67. Gunther, *Death Be Not Proud*, pp. 99–100.

68. Ruth O'Brien, *Crippled Justice: The History of Modern Disability Policy in the Workplace* (Chicago: University of Chicago Press, 2001), p. 30.

69. Gunther, *Death Be Not Proud*, p. 77.

70. Dushkin, Dec. 20, 1962, DSD.

71. See Nancy Tomes, "Celebrity Diseases," in *Medicine's Moving Pictures: Medicine, Health, and Bodies in American Film and Television*, ed. Leslie J. Reagan, Nancy Tomes, and Paula A. Treichler (Rochester: University of Rochester Press, 2007), p. 58; see also Barron H. Lerner, *When Illness Goes Public: Celebrity Patients and How We Look at Medicine* (Baltimore: Johns Hopkins University Press, 2006).

72. Deane Heller and David Heller, *John Foster Dulles: Soldier for Peace* (New York: Holt, Rinehart and Winston, 1960).

73. James T. Patterson, *Mr. Republican: A Biography of Robert A. Taft* (Boston: Houghton Mifflin, 1972); Jhan Robbins and June Robbins, *Eight Weeks to Live: The Last Chapter in the Life of Senator Robert A. Taft* (Garden City, NY: Doubleday, 1954), p. 6.

74. Heller and Heller, *John Foster Dulles*, p. 303, Robbins and Robbins, *Eight Weeks to Live*, p. 13.

75. "Mrs. Zaharias Resting," *New York Times*, Aug. 7, 1955.

76. "Obituary: Babe Zaharias Dies; Athlete Had Cancer," *New York Times*, Sept. 28, 1956.
77. "Mr. Dulles' Illness," *New York Times*, Feb. 15, 1959.
78. "Mrs. Zaharias Expected to Return to Golf Play," *New York Times*, Sept. 11, 1955.
79. Heller and Heller, *John Foster Dulles*, pp. 294–95.
80. Robbins and Robbins, *Eight Weeks to Live*, pp. 14–15.
81. "Mrs. Zaharais 'Losing Ground,'" *New York Times*, Sept. 23, 1956.
82. Heller and Heller, *John Foster Dulles*, p. 295.
83. Ibid., p. 195.
84. Robbins and Robbins, *Eight Weeks to Live*, p. 22.
85. "Cancer Fund Set Up by Babe Didrikson," *New York Times*, Sept. 13, 1955; "Birthday Party Cheers Mrs. Zaharias, 42, as She Pleads for Cancer Research Fund," *New York Times*, June 27, 1956.
86. Howard A. Rusk, "Cancer, Dulles and Hope," *New York Times*, Feb. 22, 1959.
87. "Taft Cancer Fund Urged," *New York Times*, Aug. 2, 1953.
88. "President Vows to Lead a Drive against Cancer," *New York Times*, May 12, 1971.
89. See Starr, *Social Transformation of American Medicine*; Rosemary Stevens, *In Sickness and In Wealth: American Hospitals in the Twentieth Century* (New York: Basic Books, 1989).
90. Dushkin Diary, Dec. 6, 1959, DSD.

Chapter 6 • "Nothing More to Do"

1. O. G. Brim, H. E. Freeman, S. Levine, and N. A. Scotch, *The Dying Patient* (New York: Russell Sage Foundation, 1970); Rosemary Stevens, *In Sickness and In Wealth: American Hospitals in the Twentieth Century* (New York: Basic Books, 1989), p. 231.
2. Judith R. Lave and Lester B. Lave, *The Hospital Construction Act: An Evaluation of the Hill-Burton Program, 1948–1973*, Evaluation Studies 16 (Washington, DC: American Enterprise Institute for Public Policy Research, 1974), pp. 7–15.
3. Health Insurance Institute, *Source Book of Health Insurance Data, 1963* (New York: Health Insurance Institute, 1963), p. 11; see also Julie Fairman and Joan E. Lynaugh, *Critical Care Nursing: A History* (Philadelphia: University of Pennsylvania Press, 1998), pp. 27–28.
4. AMA, *Report of the Commission on the Cost of Medical Care* (Washington, DC: American Medical Association, 1964), p. 142; Fairman and Lynaugh, *Critical Care Nursing*, pp. 29, 67; Stevens, *In Sickness and In Wealth*, p. 231.
5. A 1965 study reported that approximately 30 percent of patients hospitalized for myocardial infarction died. See Clarence A. Imboden, Jr., and Jane E. Wynn, "The Coronary Care Area," *American Journal of Nursing* 65, no. 2 (Feb. 1965): 72.
6. See Fairman and Lynaugh, *Critical Care Nursing*.
7. Margaret Nestor to Sigmund Nestor, Oct. 31 and Nov. 3, 1942, files 6 and 7, box 4, SMN.
8. George Barrett, "Making the Rounds with a Nurse," *New York Times*, Nov. 15, 1959.
9. See Leonard Engel, "The Ills of 'Maintown' Hospital," *New York Times*, Nov. 26, 1961; Martin Tolchin, "The Nurses' Gains," *New York Times*, May 28, 1966. See also Robert E. Bulander, Jr., "'The Most Important Problem in the Hospital': Nursing in the Development of the Intensive Care Unit, 1950–1965," *Social History of Medicine* 23, no. 3 (2010): 621–38.

10. "Changes in Nursing from the 1930's to 1953," *Nursing Research* 5, no. 2 (Oct. 1956): 85–86. See also Julie Fairman and Sarah Kagan, "Creating Critical Care: The Case of the Hospital of the University of Pennsylvania, 1950–1965," *Advances in Nursing Science* 22, no. 1 (Sept. 1999): 63–77.

11. Fairman and Lynaugh, *Critical Care Nursing*, p. 21.

12. Letter of Thomas Thompson Trustees, Oct. 1, 1952, Brattleboro Cases, file no. 10, box 77, TTT.

13. See Julie Fairman, "Watchful Vigilance: Nursing Care, Technology, and the Development of Intensive Care Units," *Nursing Research* 41, no. 1 (Jan./Feb. 1992): 56–57.

14. Barrett, "Making the Rounds."

15. Fairman and Lynaugh, *Critical Care Nursing*, p. 2.

16. Ibid., p. 114.

17. Jane Barton, "Round the Clock Nursing or Self-Serve: Patient Care Is Based on Medical Need," *Modern Hospital* 88 (June 1957): 51. See also Max S. Sadove, James Cross, Harry G. Higgins, and Manuel J. Segall, "The Recovery Room Expands Its Service," *Modern Hospital* 83, no. 5 (Nov. 1954): 66.

18. P. Safar, T. J. DeKornfeld, J. W. Pearson, and J. S. Redding, "The Intensive Care Unit: A Three Year Experience at Baltimore City Hospitals," *Anaesthesia* 16, no. 3 (July 1961): 278.

19. William T. Mosenthal and David D. Boyd, "Special Unit Saves Lives, Nurses, and Money," *Modern Hospital* 88, no. 6 (Dec. 1957): 86, 84. See also Barney G. Glaser and Anselm L. Strauss, *Awareness of Dying* (New Brunswick, NJ: Aldine Transactions, 1965), p. 178; Barney G. Glaser and Anselm L. Strauss, *Time for Dying* (New Brunswick, NJ: Aldine Transactions, 1968), p. 41.

20. Glaser and Strauss, *Awareness of Dying*.

21. Imboden and Wynn, "Coronary Care Area," pp. 72–76; Rose Pinneo, "Nursing in a Coronary Care Unit," *American Journal of Nursing* 65, no. 2 (Feb. 1965): 76–79.

22. Community Hospital of Battle Creek, *The Planning and Operation of an Intensive Care Unit: An Experience Brochure* (Battle Creek, MI: W. K. Kellogg Foundation, [1960]), p. 22.

23. Mosenthal and Boyd, "Special Unit," p. 86.

24. Fairman and Lynaugh, *Critical Care Nursing*, p. 17.

25. See Walter C. Alvarez, "Care of the Dying," *JAMA* 150, no. 2 (Sept. 13, 1952): 86–91; O. R. Bowen, "Why Cancer Victims Should Be Told the Truth," *Medical Times* 38, no. 8 (1955): 793–99; William D. Kelly and Stanley R. Friesen, "Do Cancer Patients Want to Be Told?" *Surgery* 27, no. 6 (1950): 822–26; F. G. H. Maloney, "Should a Patient Be Told He Has Cancer?" *Wisconsin Medical Journal* 53 (1954): 541–44; Samuel Standard and Helmuth Nathan, eds., *Should the Patient Know the Truth? A Response of Physicians, Nurses, Clergymen, and Lawyers* (New York: Springer, 1955). See also Mary Ann Krisman-Scott, "An Historical Analysis of Disclosure of Terminal Status," *Journal of Nursing Scholarship*, First Quarter 2000, pp. 47–52.

26. "Way of Dying," *Atlantic Monthly*, Jan. 1957, p. 53.

27. "A New Way of Dying," *Reader's Digest*, Mar. 1957.

28. "Life-in-Death," *New England Journal of Medicine* 256 (Apr. 18, 1957): 760.

29. Frank J. Ayd, Jr., "The Hopeless Case: Medical and Moral Considerations," *JAMA* 181, no. 13 (Sept. 19, 1962): 1100. For other works by physicians in the early 1960s question-

ing the tendency to prolong life when the possibility of recovery dims, see Thomas T. Jones, "Dignity in Death: The Application and Withholding of Interventive Measures," *Journal of the Louisiana State Medical Society* 113, no. 5 (May 1961): 180–83; James A. Knight, "Philosophic Implications of Terminal Illness," *North Carolina Medical Journal*, Oct. 1961, pp. 493–95.

30. Herman Feifel, ed., *The Meaning of Death* (New York: McGraw-Hill, 1959).

31. Thomas D. Eliot, review of *The Meaning of Death*, ed. Herman Feifel, *Sociological Quarterly* 4, no. 1 (Winter 1963): 92–93.

32. Howard S. Becker, Blanche Geer, Everett C. Hughes, and Anselm L. Strauss, *Boys in White: Student Culture in Medical School* (Chicago: University of Chicago Press, 1961).

33. Glaser and Strauss, *Awareness of Dying*; Glaser and Strauss, *Time for Dying*; Jeanne C. Quint, *The Nurse and the Dying Patient* (New York: Macmillan, 1967); David Sudnow, *Passing On* (Englewood Cliffs, NJ: Prentice-Hall, 1967).

34. Becker et al., *Boys in White*, pp. 316, 228, 317.

35. Quint, *The Nurse and the Dying Patient*, p. 20.

36. Glaser and Strauss, *Time for Dying*, pp. 46–47.

37. Glaser and Strauss, *Awareness of Dying*, p. 205.

38. Ibid., pp. 177–225.

39. Barney G. Glaser and Anselm L. Strauss, "The Social Loss of Dying Patients," *American Journal of Nursing* 64, no. 6 (June 1964): 121.

40. Dushkin Diary, Jan. 5, 1962, DSD.

41. Glaser and Strauss, "Social Loss of Dying Patients," p. 120.

42. Sudnow, *Passing On*, p. 25.

43. Ibid., p. 65.

44. Glaser and Strauss, *Awareness of Dying*, p. 146.

45. See Avery D. Weisman and J. William Worden, "The Social Significance of the Danger List," *JAMA* 215, no. 12 (Mar. 22, 1971): 1963–66.

46. Sudnow, *Passing On*, pp. 78, 80, 139, 78.

47. See Donald M. Berwick and Meera Kotagal, "Restricted Visiting Hours in the ICUs: Time to Change," *JAMA* 292, no. 6 (2004): 736–37.

48. Robert V. Wells, *Facing the "King of Terrors": Death and Society in an American Community, 1750–1990* (New York: Cambridge University Press, 2000), p. 191.

49. Dushkin Diary, Jan. 5, 1962, DSD.

50. Graham Mooney and Jonathan Reinarz, "Hospital and Asylum Visiting in Historical Perspective," in *Permeable Walls: Historical Perspectives on Hospital and Asylum Visiting*, ed. Graham Mooney and Jonathan Reinarz (New York: Editions Rodopi, 2009), p. 10.

51. The two patient names in this paragraph and the next are pseudonyms.

52. PHPR #715199.

53. PHPR #725236.

54. H.M. to John Gunther, Feb. 16. 1949, folder 1, box 47, John Gunther Papers, Special Collections Research Library, University of Chicago Library.

55. Interview with Joan Lynaugh, Oct. 19, 2010; Barrett, "Making the Rounds."

56. Margaret Nestor to Sigmund Nestor, Oct. 23, 1942, file 6, box 4, SMN.

57. Glaser and Strauss, *Awareness of Dying*, pp. 158–59.

58. Glaser and Strauss, *Time for Dying*, p. 14.

59. "Thanatology: Death and Modern Man," *Time*, Nov. 20, 1964.

60. Glaser and Strauss, *Time for Dying*, pp. 198–99.

61. Nancy Kline, "Florida," *Persimmon Tree Magazine*, Summer 2010; see also Glaser and Strauss, *Time for Dying*, pp. 198–99.

62. Max S. Sadove, James Cross, Harry G. Higgins, and Manuel J. Segall, "The Recovery Room Expands Its Service," *Modern Hospital* 83, no. 5 (Nov. 1954): 70.

63. See Jenny B. Hammer, "Visitation Policies in the ICU: A Time for Change," *Critical Care Nurse* 10, no. 1 (1990): 48; Elisabeth Kübler-Ross, "What Is It Like to Be Dying?" *American Journal of Nursing* 71, no. 1 (Jan. 1971): 55.

64. Ramona Davidson, "A Medical Intensive Care Unit," *American Journal of Nursing* 64, no. 12 (Dec. 1964): 80; Sadove et al., "Recovery Room," p. 70.

65. Elisabeth Kübler-Ross, "Dying with Dignity," *Canadian Nurse* 67, no. 10 (Oct. 1971): 35.

66. John Gunther, *Death Be Not Proud* (New York: Harper & Row, 1949), p. 137; Kline, "Florida"; Jones, "Dignity in Death," p. 181.

67. See *The Human Radiation Experiments: Final Report of the President's Advisory Committee* (New York: Oxford University Press, 1996), pp. 6, 7, 14, 227–62.

68. Quoted in Gilbert Whittemore and Miriam Boleyn-Fitzgerald, "Injecting Comatose Patients with Uranium: America's Overlapping Wars against Communism and Cancer in the 1950s," in *Useful Bodies: Humans in the Service of Medical Science in the Twentieth Century*, ed. Jordan Goodman, Anthony McElligott, and Lara Marks (Baltimore: Johns Hopkins University Press, 2003), p. 180.

69. Ibid., p. 171.

70. *Human Radiation Experiments*, p. 83.

71. "Addendum to Request Submitted to the American Cancer Society for Aid to the Cancer Chemotherapy Program of the Memorial Hospital," box 93, 1947, ML.

72. *Human Radiation Experiments*, p. 253.

73. "Ground Is Broken at Ewing Hospital," *New York Times*, Jan. 25, 1947; "Delafield Hospital Will Treat Cancer," *New York Times*, May 27, 1951; "City Set to Open New Cancer Unit," *New York Times*, Aug. 28, 1950; "New Cancer Unit Dedicated by City," *New York Times*, Aug. 31, 1950.

74. *James Ewing Hospital, Department of Hospitals, City of New York and Memorial Center for Cancer and Allied Diseases* (n.p., 1950), p. 6, file 74, box 4, Hayes Martin Collection, Records of the Memorial Sloan Kettering Cancer Center, Rockefeller Archives, Sleepy Hollow, NY.

75. Department of Hospitals, City of New York, *Annual Report, 1950* (New York: Department of Hospitals, 1950), p. 18.

76. "Delafield Hospital Will Treat Cancer"; "City Set to Open New Cancer Unit."

77. Memorandum from John H. Teeter to Research Committee, Sept. 23, 1949, folder "American Cancer Society 1949," box 95, ML.

78. C. P. Rhoads to D. C. Morris Jacobs, June 25, 1957, folder "Memorial Hospital-SKI," box 116, ML.

79. "A Plan for Accelerated Access to the Results of Cancer Research for Patients in the New York City Hospital System," enclosed in letter from C. P. Rhoads to Mary Lasker, Nov. 25, 1957, folder "Memorial Hospital-SKI," box 116, ML.

80. "Nurse Lack Acute at City Hospitals," *New York Times*, Mar. 4, 1952.

81. Department of Hospitals, New York City, *Annual Report, 1950*, p. 29.

82. "Nurse Lack Acute."

83. C. P. Rhoads to Morris A. Jacobs, Sept. 17, 1957, folder "Memorial Hospital-SKI," box 116, ML.

84. Willard C. Rappleye to James S. Adams, Oct. 27, 1947, folder "Cancer Material," box 94, ML; C. P. Rhoads to Mary Lasker, May 22, 1952, folder "Memorial Hospital, 1952," box 115, ML; Rhoads to Morris A. Jacobs, Sept. 17, 1957, folder "Memorial Hospital," box 116, ML.

85. C. P. Rhoads to Dr. Leona Baumgartner, Dec. 18, 1956, folder "Memorial Hospital," box 116, ML.

86. Sloan-Kettering Institute for Cancer Research, *Biennial Report, July 1, 1955–June 30, 1957* (New York: Sloan Kettering Institute for Cancer Research, 1957), pp. 8–9.

87. Morris A. Jacobs to Abraham D. Beame, Jan. 6, 1958, folder "Memorial Hospital," box 116, ML.

88. "Cancer-Fighter Cornelius P. Rhoads," *Time*, June 27, 1949.

89. Quoted in Susan E. Lederer, "'Porto Ricochet': Joking about Germs, Cancer, and Race Extermination in the 1930s," *American Literary History* 14, no. 4 (Winter 2002): 720–46.

90. Ibid.

91. See Eric T. Rosenthal, "The Rhoads Not Given: The Tainting of the Cornelius P. Rhoads Memorial Award," *Oncology Times* 25, no. 17 (Sept. 10, 2003): 19–20.

92. *Human Radiation Experiments*, p. 83.

93. David Pacchioli, "Subjected to Science," *Research/Penn State* 17, no. 1 (Mar. 1996).

94. "Special Ward Aids Cancer Research," *New York Times*, June 3, 1952.

95. Quotations from *Human Radiation Experiments*, pp. 83–84.

96. Gerald Kutcher, *Contested Medicine: Cancer Research and the Military* (Chicago: University of Chicago Press, 2009).

97. Ilana Löwy, *Between Bench and Bedside: Science, Healing, and Interleukin-2 in a Cancer Ward* (Cambridge: Harvard University Press, 1996), pp. 54–84.

98. Rhoads used the phrase "far advanced" in a July 17, 1953, letter to Mary Lasker describing the chemotherapy research ward recently established at Ewing Hospital, folder "Memorial Hospital," box 115, ML.

99. "Suggestion for the Organization of the Francis Delafield Hospital," file "Francis Delafield Hospital, 1950," box 319, Columbia University Office of the VP for Health Sciences, Central Records, Archives and Special Collections, Columbia University Health Sciences Library, New York.

100. Earl Ubell, "City Cancer Hospital is 25% Empty," *New York Herald Tribune*, Feb. 22, 1954.

101. Marta Fraenkel to Henry Kolbe, Sept. 5, 1957, file "Francis Delafield Hospital, 1956–1959," box 320, Columbia University Office of the VP for Health Sciences, Central Records.

102. C. P. Rhoads to Morris Jacobs, June 25, 1957, folder "Memorial Hospital," box 116, ML.

Chapter 7 • A Place to Die

1. Stephen Novak and his assistants at the Archives and Special Collections, Augustus C. Long Health Sciences Library, Columbia University, saved twenty-nine of the records used in

this section, which date primarily from the early and middle 1940s. Historian Barron H. Lerner saved the other twenty records; most date from the late 1950s and the early 1960s.

2. Ruth Abrams, Gertrude Jameson, Mary Poehlman, and Sylvia Snyder, "Terminal Care in Cancer: A Study of Two Hundred Patients Attending Boston Clinics, *New England Journal of Medicine* 232, no. 25 (June 21, 1945): 720.

3. Institute of Medicine of Chicago, *Terminal Care for Cancer Patients: A Survey of the Facilities and Services Available and Needed for the Terminal Care of Cancer Patients in the Chicago Area* (Chicago: Central Service for the Chronically Ill of the Institute of Medicine of Chicago, 1950), p. 38. The quotation is from Ernest M. Daland, "Palliative Treatment of the Patient with Advanced Cancer," *JAMA* 136, no. 6 (Feb. 8, 1948): 391.

4. Institute of Medicine, *Terminal Care*, p. 109.

5. Ibid., pp. 37, 52.

6. Ibid., pp. 11, 76, 81.

7. Ibid., p. 134.

8. Audrey W. M. Ward, "Terminal Care in Malignant Disease," *Social Science and Medicine* 8, no. 7 (1974): 413.

9. Institute of Medicine, *Terminal Care*, p. 135.

10. Ibid., p. 19.

11. Ibid., p. 27.

12. Minnie Radamacher to Mary Post Zimmerman, June 5, 1915, file 12, box 3, Zimmerman Family Papers, HL.

13. Institute of Medicine, *Terminal Care*, p. 14.

14. Ibid., p. 135.

15. Ibid., p. 136.

16. Ibid., p. 158.

17. Ibid., pp. 41, 45.

18. Columbia-Presbyterian Medical Center, *Seventy-Seventh Annual Report, The Presbyterian Hospital, including Babies Hospital, Neurological Institute, New York Orthopaedic Hospital, Sloane Hospital, Vanderbilt Clinic, for the Twelve Months Ending December 31, 1945*, p. 1; see also Kenneth M. Ludmerer, *Learning to Heal: The Development of Medical Education* (Baltimore: Johns Hopkins University Press, 1996), p. 220.

19. Interview with Stephen Novak, Sept. 13, 2011.

20. Social Service, "Annual Report of 1963," p. 1, in Elizabeth R. Pritchard, "History of Social Service of Presbyterian Hospital" (typescript, n.d.), Archives and Special Collections, Augustus C. Long Health Sciences Library, Columbia University.

21. Social Service, "Annual Report of 1958," p. 1, in Pritchard, "History."

22. Pritchard, "History," p. 1.

23. Social Service, "Annual Report of 1957," p. 1, in Pritchard, "History."

24. PHPR #766642. Pseudonyms are used for all patient names in this chapter.

25. PHPR #317371.

26. Kenneth M. Ludmerer, *Time to Heal: American Medical Education from the Turn of the Century to the Era of Managed Care* (New York: Oxford University Press, 1999), pp. 119–20, 165–66; Alice Ullmann and Gene G. Kassebaum, "Referrals and Services in a Medical Social Work Department," *Social Service Review* 35, no. 3 (Sept. 1961): 258–67.

27. Ludmerer, *Learning to Heal*, p. 120.

28. PHPR #394039.

29. PHPR #710258.
30. PHPR #665083.
31. PHPR #703169.
32. PHPR #394039.
33. PHPR #396389.
34. PHPR #752632.
35. PHPR #707009.
36. PHPR #684049.
37. PHPR #726014.
38. PHPR #706640.
39. PHPR #715199.
40. PHPR #695182.
41. PHPR #714385.
42. PHPR #781886.
43. PHPR #781886.
44. PHPR #710258.
45. Abrams et al., "Terminal Care in Cancer," p. 723.
46. Social Service, "Annual Report of 1954," p. 13, in Pritchard, "History."
47. PHPR #394039.
48. PHPR #396128.
49. PHPR #706640.
50. Bruce C. Vladeck, *Unloving Care: The Nursing Home Tragedy* (New York: Basic Books, 1980).
51. Ellen Schell, "The Origins of Geriatric Nursing: The Chronically Ill Elderly in Almshouses and Nursing Homes, 1900–1950," *Nursing History Review* 1 (1993): 203–16.
52. Theda L. Waterman, "Nursing Homes—Are They Homes? Is There Nursing?" *American Journal of Public Health* 43, no. 3 (1953): 308.
53. See "Nursing Home Care in Maryland," *Public Health Reports* 77, no. 1 (Jan. 1962): 89–90; Ollie A. Randall, "The Situation with Nursing Homes," *American Journal of Nursing* 65, no. 11 (Nov. 1965): 92–97; Schell, "The Origins of Geriatric Nursing"; Jerry Solon, "On Patients and Their Care: Proprietary Nursing Homes," *Public Health Reports* 71, no. 7 (July 1956): 646–51; Waterman, "Nursing Homes," pp. 307–13.
54. PHPR #706640.
55. Social Service, "Annual Report of 1964–65," p. 3, in Pritchard, "History."
56. PHPR #763015.
57. PHPR #71900.
58. PHPR #456638.
59. PHPR #763015.
60. PHPR #71900.
61. PHPR #706640.
62. PHPR #396128.
63. PHPR #671660.
64. PHPR #307462.
65. PHPR #306872.
66. PHPR #394039.
67. PHPR #703621.

68. PHPR #665083.
69. PHPR #695182.
70. PHPR #665083.
71. PHPR #257751.
72. Abrams et al., "Terminal Care in Cancer," p. 720.
73. PHPR #715199.
74. Philippe Ariès, *The Hour of Our Death*, trans. Helen Weaver (New York: Vintage, 1981), pp. 570–71.
75. Institute of Medicine of Chicago, *Terminal Care*, p. 128.
76. Ibid., p. 18.
77. Abrams et al., "Terminal Care in Cancer," p. 721.
78. PHPR #707009.
79. PHPR #701614.
80. PHPR #665083.

Chapter 8 · The Sacred and the Spiritual

1. See Ira Byock, *The Best Care Possible: A Physician's Quest to Transform Care through the End of Life* (New York: Avery, 2012); Barbara Clow, *Negotiating Disease: Power and Cancer Care, 1900–1950* (Montreal: McGill-Queen's University Press, 2001); Robert Fulton, "Death and the Self," *Journal of Religion and Health* 3, no. 4 (July 1964): 363.
2. See Jon Butler, Grant Wacker, and Randall Balmer, *Religion in American Life: A Short History* (New York: Oxford University Press, 2008); Richard Hughes Seager, *Buddhism in America* (New York: Columbia University Press, 1999).
3. Dushkin Diary, Mar. 30, 1961, DSD.
4. Dushkin Diary, Feb. 17, 1960.
5. Dushkin Diary, Feb. 27, 1961.
6. Dushkin Diary, Jan. 5, 1962.
7. Quoted in Gretchen Krueger, *Hope and Suffering: Children, Cancer, and the Paradox of Experimental Medicine* (Baltimore: Johns Hopkins University Press, 2008), p. 58.
8. Frances Gunther, "A Word from Frances," in *Death Be Not Proud*, by John Gunther (New York: Harper and Row, 1949), p. 190.
9. Ibid., p. 193.
10. See Lucy Bregman and Sara Thiermann, *First Person Mortal: Personal Narratives of Illness, Dying, and Grief* (New York: Paragon House, 1995).
11. Gunther, "A Word from Frances," pp. 188–89.
12. Ibid., p. 191.
13. Ibid., pp. 194–95.
14. Letters to Frances Gunther from Lillian L. Lieber, Feb. 24, 1949, folder 58; Dora S. Raue, Mar. 22, 1949, folder 59; Irene Gromik, Feb. 21, 1949, folder 58; "A Mother," Jan. 26, 1949, folder 58. All letters to Gunther are in box 2 of the Frances Fineman Gunther Papers, 1915–63, Arthur and Elizabeth Schlesinger Library, Radcliffe College, Cambridge, MA.
15. Letters to Gunther from Mrs. Willard Van Hazan, Mar. 3, 1949, folder 59; Sophie A. Udin, Mar. 5, 1949, folder 59; Mrs. C. E. Morrison, Apr. 4, 1949, folder 59.
16. Letters to Gunther from Lora J. Jackson, Mar. 1, 1949, folder 59; Georgiane A. Dillard, Feb. 23, 1949, folder 58; Marianne R. Peters, Feb. 13, 1949, folder 58.

17. Letters to Gunther from Eileen Bowen, Apr. 10, 1949, folder 59; Elizabeth Custer Nearing, Feb. 21, 1949, folder 58.

18. Barney G. Glaser and Anselm L. Strauss, *Awareness of Dying* (Chicago: Aldine, 1965), p. 127.

19. Renée C. Fox, *Experiment Perilous: Physicians and Patients Facing the Unknown* (1959; New Brunswick, NJ: Transaction, 1998), pp. 177–78.

20. Robert V. Wells, *Facing the "King of Terrors": Death and Society in an American Community, 1750–1990* (New York: Cambridge University Press, 2000), p. 236.

21. Ibid., pp. 190–91.

22. The most notable recent study is in Robert A. Aronowitz, *Unnatural History: Breast Cancer and American Society* (New York: Cambridge University Press, 2007), pp. 183–209. Other major accounts are Linda Lear, *Rachel Carson: Witness for Nature* (New York: Henry Holt, 1997), an excellent and extremely comprehensive biography, and Ellen Leopold, *A Darker Ribbon: Breast Cancer, Women, and Their Doctors in the Twentieth Century* (New York: Beacon, 1999), which includes many letters between Carson and her primary doctor. My discussion of Flannery O'Connor relies heavily on Brad Gooch, *Flannery: A Life of Flannery O'Connor* (New York Little, Brown, 2009). Other studies that discuss her illness include Jean W. Cash, *Flannery O'Connor: A Life* (Knoxville: University of Tennessee Press, 2002); Alice Little Caldwell, "Flannery O'Connor: Life with Lupus," *Journal of the Medical Association of Georgia* 92, no. 2 (Summer 2003): 15–17; Robert Coles, *Flannery O'Connor's South* (Baton Rouge: Louisiana State University Press, 1980); Susanna Gilbert, "'Blood Don't Lie': The Diseased Family in Flanner O'Connor's *Everything That Rises Must Converge*," *Literature and Medicine* 18, no. 1 (1999): 114–31; Josephine Hendin, *The World of Flannery O'Connor* (Bloomington: Indiana University Press, 1970); Kathleen Spaltro, "When We Dead Awaken: Flannery O'Connor's Debt to Lupus," *Flannery O'Connor Bulletin* 20 (1991): 33–44; Sue Walker, "Spelling Out Illness: Lupus as Metaphor in Flannery O'Connor's 'Greenleaf,'" *Chattahoochee Review* 1 (1991): 54–63.

23. See especially Gilbert, "'Blood Don't Lie'"; Spaltro, "When We Dead Awaken"; Walker, "Spelling Out Illness."

24. Rita Charon, *Narrative Medicine: Honoring the Stories of Illness* (New York: Oxford University Press, 2006), p. 65.

25. Quoted in Gooch, *Flannery: A Life*, pp. 46–47.

26. Ibid., p. 69.

27. Quoted ibid., p. 72.

28. Quoted ibid., p. 185.

29. "The Enduring Chill," in *Flannery O'Connor: Collected Works*, ed. Sally Fitzgerald (New York: Library of America, 1988), p. 547.

30. See Daniel M. Fox, *Power and Illness: The Failure and Future of American Health Policy* (Berkeley: University of California Press, 1993); see also Harry M. Marks, "Cortisone, 1949: A Year in the Political Life of a Drug," *Bulletin of the History of Medicine* 66, no. 3 (1992): 419–39.

31. George W. Thorne, "Development and Application of ACTH and Cortisone," *Bulletin of the American Academy of Arts and Sciences* 4, no. 5 (Feb. 1951): 4.

32. Quoted Gooch, *Flannery: A Life*, p. 193.

33. *The Habit of Being: Letters of Flannery O'Connor*, ed. Sally Fitzgerald (New York: Farrar, Straus and Giroux, 1979), p. 22.

34. Quoted in Gooch, *Flannery: A Life*, p. 216.
35. Quoted ibid., p. 217.
36. See *Habit of Being*, pp. 67, 107.
37. Ibid., p. 57.
38. "The Life You Save May Be Your Own," in *Flannery O'Connor: Collected Works*, p. 173, quoted in Gooch, *Flannery: A Life*, p. 217.
39. Joyce Carol Oates, "The Parables of Flannery O'Connor," *New York Review of Books*, Apr. 9, 2009.
40. See, e.g., Cash, *Flannery O'Connor*, pp. 147–73; Gooch, *Flannery: A Life*, pp. 317, 341.
41. Rebecca Jo Plant, *Mom: The Transformation of Motherhood in Modern America* (Chicago: University of Chicago Press, 2010), p. 19.
42. See ibid., pp. 86–117.
43. "The Comforts of Home," in *Flannery O'Connor: Collected Works*, p. 573.
44. See Gilbert, " Blood Don't Lie.'" I also am grateful for the insights of Harold Braswell. The stories can be found in *Flannery O'Connor: Collected Works*.
45. Gooch, *Flannery: A Life*, pp. 223, 228.
46. *Habit of Being*, p. 114.
47. Quoted in Gooch, *Flannery: A Life*, pp. 270–71.
48. *Habit of Being*, p. 96.
49. Ibid., p. 163.
50. Carol Shloss, "'Cheers, Tarfunk': The Letters of Flannery O'Connor," *Massachusetts Review* 20, no. 2 (1979): 388.
51. *Habit of Being*, pp. 106, 107.
52. "Good Country People," in *Flannery O'Connor: Collected Works*, p. 275.
53. Quoted in Gilbert, "'Blood Don't Lie,'" p. 118.
54. Quoted in Gooch, *Flannery: A Life*, p. 301.
55. Ibid., p. 306.
56. *Habit of Being*, p. 322.
57. "God-Intoxicated Hillbillies," *Time*, Feb. 29, 1960, p. 118.
58. *Habit of Being*, pp. 378, 380.
59. Ibid., p. 512.
60. Ibid., p. 509.
61. Ibid., p. 328.
62. Ibid., p. 409.
63. The introduction is reprinted in Flannery O'Connor, *Mystery and Manners: Occasional Prose*, ed. Sally Fitzgerald and Robert Fitzgerald (New York: Farrar, Straus, and Giroux, 1957); the quotation appears on p. 223.
64. *Habit of Being*, p. 420.
65. Ibid., p. 423.
66. Ibid., p. 441.
67. Ibid., p. 462.
68. Ibid., p. 521.
69. Ibid., p. 549.
70. Ibid., p. 560.
71. Ibid., p. 560.
72. Gooch, *Flannery: A Life*, p. 359.

73. *Habit of Being*, p. 587.

74. Ibid., p. 591.

75. See Robert Wuthnow, *After Heaven: Spirituality in America since the 1950s* (Berkeley: University of California Press, 1998).

76. Rachel Carson, "Clouds," in *Lost Woods: The Discovered Writing of Rachel Carson*, ed. Linda Lear (Boston: Beacon Press, 1998), p. 178.

77. Rachel Carson, *The Edge of the Sea* (New York: Times-Mirror / New American Library, 1955), p. 11

78. Rachel Carson, Preface to the Revised Edition, *The Sea Around Us* (New York: Times-Mirror / New American Library, 1961), p. ix.

79. Rachel Carson, "The Real World around Us," in Lear, *Lost Woods*, p. 163

80. Rachel Carson, "Remarks at the Acceptance of the National Book Award for Nonfiction," in Lear, *Lost Woods*, p. 89.

81. Rachel Carson, *The Sense of Wonder* (New York: Harper Collins, 1998), p. 54.

82. Lear, *Rachel Carson*, p. 16.

83. Ibid., pp. 21–23

84. Brooks, *The House of Life: Rachel Carson at Work* (Boston: Houghton Mifflin, 1972), pp. 19–20.

85. *Always, Rachel: The Letters of Rachel Carson and Dorothy Freeman, 1952–1964; The Story of a Remarkable Friendship*, ed. Martha Freeman (Boston: Beacon Press, 1995), p. 49.

86. Lear, *Rachel Carson*, p. 136.

87. *Always, Rachel*, p. 23.

88. See Paula DiPerna, "The Conscience of the Age," *Women's Review of Books* 15, no. 6 (Mar. 1998): 1, 3–4.

89. Rachel's will stipulated that Dorothy should receive the letters she had written. Before her death, Dorothy gave both sets of letters to her granddaughter, who published them in 1995. See *Always, Rachel*. This volume does not include the entire correspondence. Some letters were not preserved, and some, especially those from the early years of the relationship, were excluded by the editor. Dorothy's side of the exchange is especially scanty. Nevertheless, the volume includes the great majority of letters Carson wrote during the last three years of her life, when she contended with breast cancer. See the Introduction to *Always, Rachel*, p. xix.

90. *Always, Rachel*, p. 98.

91. Brooks, *House of Life*, p. 197.

92. *Always, Rachel*, p. 273.

93. Quoted in Lear, *Rachel Carson*, p. 365.

94. See Barron H. Lerner, *The Breast Cancer Wars: Fear, Hope, and the Pursuit of a Cure in Twentieth-Century America* (New York: Oxford University Press, 2001), pp. 15–68; James S. Olson, *Bathsheba's Breast: Women, Cancer, and History* (Baltimore: Johns Hopkins University Press, 2002), pp. 67, 89.

95. See Lerner, *Breast Cancer Wars*, pp. 69–195.

96. Quoted in Leopold, *Darker Ribbon*, p. 130.

97. Lear, *Rachel Carson*, p. 367.

98. See Rachel Carson to George Crile, Jr., Dec. 7, 1960, in Leopold, *Darker Ribbon*, p. 129

99. *Always, Rachel*, p. 313.

100. Carson to Crile, Dec. 17, 1960, in Leopold, *Darker Ribbon*, p. 129.

101. *Always, Rachel*, p. 313.

102. On the extent to which cancer patients in the 1960s attempted to participate in medical decisions, see Barron H. Lerner, "Beyond Informed Consent: Did Cancer Patients Challenge Their Physicians in the Post-World War II Era?" *Journal of the History of Medicine and Allied Sciences* 59, no. 4 (2004): 507–21.

103. Quoted in Brooks, *House of Life*, p. 265.

104. George Crile, *Cancer and Common Sense* (New York: Viking, 1955), p. 4.

105. Rachel Carson, *Silent Spring*, 40th Anniversary Edition (Boston: Houghton Mifflin, 2002), p. 296.

106. Crile, *Cancer and Common Sense*, pp. 83, 105.

107. Carson to Crile, Dec. 17, 1960, in Leopold, *Darker Ribbon*, p. 132.

108. Crile, *Cancer and Common Sense*, p. 18.

109. Quoted in Lear, *Rachel Carson*, p. 381.

110. *Always, Rachel*, p. 325.

111. Ibid., p. 320.

112. Quoted in Lear, *Rachel Carson*, p. 381.

113. *Always, Rachel*, p. 329.

114. Ibid., p. 331.

115. Ibid., p. 337.

116. Lear, *Rachel Carson*.

117. *Always, Rachel*, p. 349.

118. Carson to Crile, Mar. 18, 1961, in Leopold, *Darker Ribbon*, p. 135.

119. *Always, Rachel*, pp. 349, 354.

120. Ibid., p. 363.

121. Ibid., p. 345.

122. Ibid., pp. 351, 352, 354, 355.

123. Ibid., p. 353.

124. Ibid., p. 360.

125. Ibid., p. 390.

126. Carson, *Silent Spring*, p. 13.

127. Ibid., p. 8.

128. Quoted in Lear, *Rachel Carson*, pp. 334–35.

129. *Always, Rachel*, p. 391.

130. Ibid., p. 394.

131. Ibid., p. 398.

132. Ibid., p. 399.

133. See Aronowitz, *Unnatural History*, p. 194.

134. *Always, Rachel*, p. 401.

135. Ibid., p. 403.

136. Ibid., pp. 403–4.

137. Ibid., p. 405.

138. Ibid., p. 407.

139. Ibid., p. 408.

140. See Michael B. Smith, "'Silence, Miss Carson!': Science, Gender, and the Reception of 'Silent Spring,'" *Feminist Studies* 27, no. 3 (Autumn 2001): 733–52.

141. *Always, Rachel*, p. 414.
142. Ibid., p. 425.
143. Carson to Crile, Feb. 17, 1963, in Leopold, *Darker Ribbon*, p. 142.
144. *Always, Rachel*, p. 426.
145. Ibid.
146. Ibid., pp. 541–42.
147. Ibid.
148. Ibid., p. 431.
149. Ibid., p. 430.
150. Ibid., p. 435.
151. Ibid., p. 434.
152. Carson to Crile, Feb. 17, 1963, in Leopold, *Darker Ribbon*, p. 142
153. *Always, Rachel*, pp. 436–37.
154. Ibid., p. 439.
155. Ibid., p. 446.
156. Ibid., pp. 446–47.
157. Ibid., p. 450.
158. Ibid., p. 452.
159. Patterson, *Dread Disease*, pp. 163–67.
160. Aronowitz, *Unnatural History*, p. 198.
161. *Always, Rachel*, p. 442.
162. Carson to Crile, Apr. 3, 1963, in Leopold, *Darker Ribbon*, p. 145.
163. *Always, Rachel*, p. 457.
164. Ibid., pp. 466–67.
165. Ibid., p. 475.
166. Ibid., pp. 482–83.
167. Ibid., p. 483.
168. Ibid., p. 499.
169. Ibid., pp. 504–5.
170. Ibid., p. 515.
171. Ibid., p. 535.
172. Karen Armstrong, *The Spiral Staircase* (New York: Alfred A. Knopf, 2004), p. 272.

Conclusion

1. Institute of Medicine, Committee on Care at the End of Life, *Approaching Death: Improving Care at the End of Life* (Washington, DC: National Academy Press, 1997), p. 17.

2. "End-of-Life Care," Dartmouth Atlas of Health Care, www.dartmouthatlas.org/keyissues/issue.aspx?con=2944, accessed Dec. 8, 2011.

3. See Steven P. Wallace, Emily K. Abel, Nadereh Pourat, and Linda Delp, "Long-Term Care and the Elderly Population," in *Changing the U.S. Health Care System: Key Issues in Health Services Policy and Management*, 3rd ed., ed. Ronald M. Andersen and Thomas H. Rice (San Francisco: Jossey-Bass, 2007), pp. 341–62.

4. G. R. Vandenbos, P. H. DeLeon, and M. S. Pallak, "An Alternative to Traditional Medical Care for the Terminally Ill: Humanitarian Policy and Political Issues in Hospice Care," *American Psychologist* 37, no. 11 (Nov. 1982): 1245.

5. Edward H. Rynearson, "You Are Standing at the Bedside of a Patient Dying of Untreatable Cancer," *CA: A Cancer Journal for Clinicians* 9, no. 3 (May/June 1959): 85.

6. See Leland Christenson, "The Physician's Role in Terminal Illness and Death" (editorial), *Minnesota Medicine* 46 (Sept. 1963): 881–82; William Kitay, "Let's Retain the Dignity of Dying," *Today's Health*, May 1966, pp. 62–69; Louis Lasagna, "Editorial: The Doctor and the Dying Patient," *Journal of Chronic Diseases* 22 (1969): 66; Samuel Standard and Helmuth Nathan, eds., *Should the Patient Know the Truth*? (New York: Springer, 1955).

7. Henry Feifel, Introduction to *The Meaning of Death*, ed. Henry Feifel (New York: McGraw-Hill, 1959), p. xiv.

8. Elisabeth Kübler-Ross, *On Death and Dying* (New York: Simon and Schuster, 1969).

9. See Larry Churchill, "The Human Experience of Dying: The Moral Primacy of Stories over Stages," *Soundings* 62, no. 1 (1979): 24–37.

10. Cited in Joy Buck, "Reweaving a Tapestry of Care: Religion, Nursing, and the Meaning of Hospice, 1945–1978," *Nursing History Review* 15 (2007): 133.

11. See David Clark, "'Total Pain,' Disciplinary Power, and the Body in the Work of Cicely Saunders, 1958–1967," *Social Science and Medicine* 49, no. 6 (1999): 733.

12. Monica Mills, interview with Florence Wald, Oral History Archive, Connecticut Women's Hall of Fame, June 10, 2003, www.cwhf.org/browse_hall/hall/people/wald.php, accessed Nov. 5, 2011.

13. Susan L. Smith and Dawn Nickel, "Nursing the Dying in Post–Second World War Canada and the United States," in *Women, Health, and Nation: Canada and the United States since 1945*, ed. Georgina Feldberg (Montreal: McGill-Queen's University Press, 2003), pp. 330–54.

14. L. H. Aiken and M. M. Marx, "Perspective on the Public Policy Debate," in *Hospice Programs and Public Policy*, ed. Paul R. Torrens (Chicago: American Hospital Association, 1985).

15. Diane E. Meier, "The Development, Status, and Future of Palliative Care," p. 18, www.rwjf.org/files/research/4588.pdf, accessed Aug. 11, 2011. See Joanne Lynn, *Sick to Death and Not Going to Take It Anymore! Reforming Health Care for the Last Years of Life* (Berkeley: University of California Press, 2004), p. 67.

16. See Kathleen M. Foley, "The Past and Future of Palliative Care," *Hastings Center Report* 35, no. 6, supplement (Nov.–Dec. 2005): s42–s46.

17. Mills, interview with Florence Wald, p. 14.

18. Joan Craven and Florence S. Wald, "Hospice Care for Dying Patients," *American Journal of Nursing* 75, no. 10 (Oct. 1975): 1821.

19. Emily K. Abel, "The Hospice Movement: Institutionalizing Innovation," *International Journal of Health Services* 16, no. 1 (1986): 71–85.

20. Ibid., p. 72.

21. See Anne Munley, *The Hospice Alternative: A New Context for Death and Dying* (New York: Basic Books, 1983).

22. Craven and Wald, "Hospice Care," p. 1822.

23. Abel, "The Hospice Movement."

24. Nitin Puri, Vinod Puri, and R. P. Dellinger, "History of Technology in the Intensive Care Unit," *Critical Care Clinics* 25 (2009): 185–200.

25. "A Controlled Trial to Improve Care for Seriously Ill Hospitalized Patients; The Study to Understand Prognoses and Preferences for Outcomes and Risks of Treatments (SUPPORT)," *JAMA* 274 (1995): 1591–98.

26. Institute of Medicine, *Approaching Death,* pp. 263–64.

27. Ibid., pp. 264–65.

28. Dartmouth Atlas of Health Care, www.dartmouthatlas.org, accessed Oct. 13, 2011. For critiques of the Dartmouth Atlas, see Peter B. Bach, "A Map to Bad Policy: Hospital Efficiency Measures in the Dartmouth Atlas," *New England Journal of Medicine* 362, no. 7 (Feb. 18, 2010): 569–70, and Michael K. Ong et al., "Looking Forward, Looking Back: Assessing Variations in Hospital Resource Use and Outcomes for Elderly Patients with Heart Failure," *Cardiovascular Quality and Outcomes* 2 (2009): 548–57.

29. Meier, "Development of Palliative Care," p. 33.

30. Ibid., p. 24; see also George J. Annas, "How We Lie," *Hastings Center Report* 25, no. 6 (1995): S12.

31. Institute of Medicine, *Approaching Death,* pp. 264–65.

32. See R. Navari, C. B. Stocking, and M. Siegler, "Preferences of Patients with Advanced Cancer for Hospice Care," *JAMA* 284 (2000): 2449.

33. David Rieff, *Swimming in a Sea of Death* (New York: Simon and Schuster, 2008), p. 150.

34. Ibid., p. 104; Meier, "Development of Palliative Care," p. 19.

35. Quoted in Molly Hennessy-Fiske, "Study Examines End-of-Life Care for Cancer Patients," *Los Angeles Times,* Nov. 16, 2010.

36. See Nancy L. Keating et al., "Physician Factors Associated with Discussions about End-of-Life Care," *Cancer,* Feb. 15, 2010, pp. 998–1006; J. W. Mack et al., "Hope and Prognostic Disclosure," *Journal of Clinical Oncology* 25, no. 35 (Dec. 2007): 5636–42.

37. Quoted in Reed Abelson, "Weighing Medical Costs of End-of-Life Care," *New York Times,* Dec. 23, 2009.

38. See, e.g., Ong et al., "Looking Forward, Looking Back."

39. See Meier, "Development of Palliative Care."

40. Lynn, *Sick to Death,* p. 42.

41. See, e.g., these Government Accountability Office reports: "Nursing Homes: Additional Steps Needed to Strengthen Enforcement of Federal Quality Standards" (1999); "Nursing Homes: More Can Be Done to Protect Residents from Abuse" (2002); "Nursing Home Fire Safety: Recent Fires Highlight Weaknesses in Federal Standards and Oversight" (2004); "Nursing Homes: Efforts to Strengthen Federal Enforcement Have Not Deterred Some Homes from Repeatedly Harming Residents" (2007). See also Charlene Harrington, Joseph T. Mullan, and Helen Carrillo, "State Nursing Home Enforcement Systems," *Journal of Health Politics, Policy and Law* 29, no. 1 (Feb. 2004): 43–73.

42. Henry J. Kaiser Family Foundation, Kaiser Commission on Medicaid and the Uninsured, "Medicaid and Long-Term Care Services and Supports," Feb. 2009, p. 1, www.kff.org/medicaid/upload/2186 06.pdf, accessed Nov. 20, 2011.

43. See Wallace et al., "Long-Term Care and the Elderly Population," pp. 341–62.

44. Enid Kassner, "Home and Community-Based Long-Term Services and Supports for Older People," Fact Sheet 222 (Washington, DC: AARP Public Policy Institute, May 2011), p. 1.

45. See Emily K. Abel, *Hearts of Wisdom: American Women Caring for Kin, 1850–1940* (Cambridge: Harvard University Press, 2000).

INDEX

Page numbers in italics refer to illustrations.